Springer Undergraduate Mathematics Series

Advisory Board

Other books in this series

A First Course in Discrete Mathematics *I. Anderson*
Analytic Methods for Partial Differential Equations *G. Evans, J. Blackledge, P. Yardley*
Applied Geometry for Computer Graphics and CAD, Second Edition *D. Marsh*
Basic Linear Algebra, Second Edition *T.S. Blyth and E.F. Robertson*
Basic Stochastic Processes *Z. Brzeźniak and T. Zastawniak*
Calculus of One Variable *K.E. Hirst*
Complex Analysis *J.M. Howie*
Elementary Differential Geometry *A. Pressley*
Elementary Number Theory *G.A. Jones and J.M. Jones*
Elements of Abstract Analysis *M. Ó Searcóid*
Elements of Logic via Numbers and Sets *D.L. Johnson*
Essential Mathematical Biology *N.F. Britton*
Essential Topology *M.D. Crossley*
Fields and Galois Theory *J.M. Howie*
Fields, Flows and Waves: An Introduction to Continuum Models *D.F. Parker*
Further Linear Algebra *T.S. Blyth and E.F. Robertson*
Game Theory: Decisions, Interaction and Evolution *J.N. Webb*
General Relativity *N.M.J. Woodhouse*
Geometry *R. Fenn*
Groups, Rings and Fields *D.A.R. Wallace*
Hyperbolic Geometry, Second Edition *J.W. Anderson*
Information and Coding Theory *G.A. Jones and J.M. Jones*
Introduction to Laplace Transforms and Fourier Series *P.P.G. Dyke*
Introduction to Lie Algebras *K. Erdmann and M.J. Wildon*
Introduction to Ring Theory *P.M. Cohn*
Introductory Mathematics: Algebra and Analysis *G. Smith*
Linear Functional Analysis, Second Edition *B.P. Rynne and M.A. Youngson*
Mathematics for Finance: An Introduction to Financial Engineering *M. Capiński and T. Zastawniak*
Matrix Groups: An Introduction to Lie Group Theory *A. Baker*
Measure, Integral and Probability, Second Edition *M. Capiński and E. Kopp*
Metric Spaces *M. Ó Searcóid*
Multivariate Calculus and Geometry, Second Edition *S. Dineen*
Numerical Methods for Partial Differential Equations *G. Evans, J. Blackledge, P. Yardley*
Probability Models *J. Haigh*
Real Analysis *J.M. Howie*
Sets, Logic and Categories *P. Cameron*
Special Relativity *N.M.J. Woodhouse*
Sturm-Liouville Theory and its Applications: *M.A. Al-Gwaiz*
Symmetries *D.L. Johnson*
Topics in Group Theory *G. Smith and O. Tabachnikova*
Vector Calculus *P.C. Matthews*
Worlds Out of Nothing: A Course in the History of Geometry in the 19th Century *J. Gray*

M.A. Al-Gwaiz

Sturm-Liouville Theory and its Applications

M.A. Al-Gwaiz
Department of Mathematics
King Saud University
Riyadh, Saudi Arabia
malgwaiz@ksu.edu.sa

Cover illustration elements reproduced by kind permission of:
Aptech Systems, Inc., Publishers of the GAUSS Mathematical and Statistical System, 23804 S.E. Kent-Kangley Road, Maple Valley, WA 98038, USA. Tel: (206) 432 - 7855 Fax (206) 432 - 7832 email: info@aptech.com URL: www.aptech.com.
American Statistical Association: Chance Vol 8 No 1, 1995 article by KS and KW Heiner 'Tree Rings of the Northern Shawangunks' page 32 fig 2.
Springer-Verlag: Mathematica in Education and Research Vol 4 Issue 3 1995 article by Roman E Maeder, Beatrice Amrhein and Oliver Gloor 'Illustrated Mathematics: Visualization of Mathematical Objects' page 9 fig 11, originally published as a CD ROM 'Illustrated Mathematics' by TELOS: ISBN 0-387-14222-3, German edition by Birkhauser: ISBN 3-7643-5100-4.
Mathematica in Education and Research Vol 4 Issue 3 1995 article by Richard J Gaylord and Kazume Nishidate 'Traffic Engineering with Cellular Automata' page 35 fig 2. Mathematica in Education and Research Vol 5 Issue 2 1996 article by Michael Trott 'The Implicitization of a Trefoil Knot' page 14.
Mathematica in Education and Research Vol 5 Issue 2 1996 article by Lee de Cola 'Coins, Trees, Bars and Bells: Simulation of the Binomial Process' page 19 fig 3. Mathematica in Education and Research Vol 5 Issue 2 1996 article by Richard Gaylord and Kazume Nishidate 'Contagious Spreading' page 33 fig 1. Mathematica in Education and Research Vol 5 Issue 2 1996 article by Joe Buhler and Stan Wagon 'Secrets of the Madelung Constant' page 50 fig 1.

Mathematics Subject Classification (2000): 34B24, 34L10

British Library Cataloguing in Publication Data
A catalogue record for this book is available from the British Library

Library of Congress Control Number: 2007938910

Springer Undergraduate Mathematics Series ISSN 1615-2085
ISBN 978-1-84628-971-2 e-ISBN 978-1-84628-972-9

Printed on acid-free paper

9 8 7 6 5 4 3 2 1

springer.com

Preface

This book is based on lecture notes which I have used over a number of years to teach a course on mathematical methods to senior undergraduate students of mathematics at King Saud University. The course is offered here as a prerequisite for taking partial differential equations in the final (fourth) year of the undergraduate program. It was initially designed to cover three main topics: special functions, Fourier series and integrals, and a brief sketch of the Sturm–Liouville problem and its solutions. Using separation of variables to solve a boundary-value problem for a second-order partial differential equation often leads to a Sturm–Liouville eigenvalue problem, and the solution set is likely to be a sequence of special functions, hence the relevance of these topics. Typically, the solution of the partial differential equation can then be represented (pointwise) by a Fourier series or a Fourier integral, depending on whether the domain is finite or infinite.

But it soon became clear that these "mathematical methods" could be developed into a more coherent and substantial course by presenting them within the more general Sturm–Liouville theory in $\mathcal{L}^2$. According to this theory, a linear second-order differential operator which is self-adjoint has an orthogonal sequence of eigenfunctions that spans $\mathcal{L}^2$. This immediately leads to the fundamental theorem of Fourier series in $\mathcal{L}^2$ as a special case in which the operator is simply d^2/dx^2. The other orthogonal functions of mathematical physics, such as the Legendre and Hermite polynomials or the Bessel functions, are similarly generated as eigenfunctions of particular differential operators. The result is a generalized version of the classical theory of Fourier series, which ties up the topics of the course mentioned above and provides a common theme for the book.

In Chapter 1 the stage is set by defining the inner product space of square integrable functions $\mathcal{L}^2$, and the basic analytical tools needed in the chapters to follow. These include the convergence properties of sequences and series of functions and the important notion of completeness of $\mathcal{L}^2$, which is defined through Cauchy sequences.

The difficulty with building Fourier analysis on the Sturm–Liouville theory is that the latter is deeply rooted in functional analysis, in particular the spectral theory of compact operators, which is beyond the scope of an undergraduate treatment such as this. We need a simpler proof of the existence and completeness of the eigenfunctions. In the case of the regular Sturm–Liouville problem, this is achieved in Chapter 2 by invoking the existence theorem for linear differential equations to construct Green's function for the Sturm–Liouville operator, and then using the Ascoli–Arzela theorem to arrive at the desired conclusions. This is covered in Sections 2.4.1 and 2.4.2 along the lines of Coddington and Levinson in [6].

Chapters 3 through 5 present special applications of the Sturm–Liouville theory. Chapter 3, which is on Fourier series, provides the prime example of a regular Sturm–Liouville problem. In this chapter the pointwise theory of Fourier series is also covered, and the classical theorem (Theorem 3.9) in this context is proved. The advantage of the $\mathcal{L}^2$ theory is already evident from the simple statement of Theorem 3.2, that a function can be represented by a Fourier series if and only if it lies in $\mathcal{L}^2$, as compared to the statement of Theorem 3.9.

In Chapters 4 and 5 we discuss some of the more important examples of a singular Sturm–Liouville problem. These lead to the orthogonal polynomials and Bessel functions which are familiar to students of science and engineering. Each chapter concludes with applications to some well-known equations of mathematical physics, including Laplace's equation, the heat equation, and the wave equation.

Chapters 6 and 7 on the Fourier and Laplace transformations are not really part of the Sturm–Liouville theory, but are included here as extensions of the Fourier series method for representing functions. These have important applications in heat transfer and signal transmission. They also allow us to solve nonhomogeneous differential equations, a subject which is not discussed in the previous chapters where the emphasis is mainly on the eigenfunctions.

The reader is assumed to be familiar with the convergence properties of sequences and series of functions, which are usually presented in advanced calculus, and with elementary ordinary differential equations. In addition, we have used some standard results of real analysis, such as the density of continuous functions in $\mathcal{L}^2$ and the Ascoli–Arzela theorem. These are used to prove the existence of eigenfunctions for the Sturm–Liouville operator in Chapter 2, and they have the advantage of avoiding any need for Lebesgue measure and integration. It is for that reason that smoothness conditions are imposed on the coefficients of the Sturm–Liouville operator, for otherwise integrability conditions would have sufficed. The only exception is the dominated convergence theorem, which is invoked in Chapter 6 to establish the continuity of the Fourier transform. This is a marginal result which lies outside the context of the Sturm–Liouville theory and could have been handled differently, but the temptation to use that powerful theorem as a shortcut was irresistible.

This book follows a strict mathematical style of presentation, but the subject is important for students of science and engineering. In these disciplines, Fourier analysis and special functions are used quite extensively for solving linear differential equations, but it is only through the Sturm–Liouville theory in $\mathcal{L}^2$ that one discovers the underlying principles which clarify why the procedure works. The theoretical treatment in Chapter 2 need not hinder students outside mathematics who may have some difficulty with the analysis. Proof of the existence and completeness of the eigenfunctions (Sections 2.4.1 and 2.4.2) may be skipped by those who are mainly interested in the results of the theory. But the operator-theoretic approach to differential equations in Hilbert space has proved extremely convenient and fruitful in quantum mechanics, where it is introduced at the undergraduate level, and it should not be avoided where it seems to brings clarity and coherence in other disciplines.

I have occasionally used the symbols $\Rightarrow$ (for "implies") and $\Leftrightarrow$ (for "if and only if") to connect mathematical statements. This is done mainly for the sake of typographical convenience and economy of expression, especially where displayed relations are involved.

A first draft of this book was written in the summer of 2005 while I was on vacation in Lebanon. I should like to thank the librarian of the American University of Beirut for allowing me to use the facilities of their library during my stay there. A number of colleagues in our department were kind enough to check the manuscript for errors and misprints, and to comment on parts of it. I am grateful to them all. Professor Saleh Elsanousi prepared the figures for the book, and my former student Mohammed Balfageh helped me to set up the software used in the SUMS Springer series. I would not have been able to complete these tasks without their help. Finally, I wish to express my deep appreciation to Karen Borthwick at Springer-Verlag for her gracious handling of all the communications leading to publication.

M.A. Al-Gwaiz
Riyadh, March 2007

Contents

1
Inner Product Space

An inner product space is the natural generalization of the Euclidean space $\mathbb{R}^n$, with its well-known topological and geometric properties. It constitutes the framework, or setting, for much of our work in this book, as it provides the appropriate mathematical structure that we need.

1.1 Vector Space

We use the symbol $\mathbb{F}$ to denote either the real number field $\mathbb{R}$ or the complex number field $\mathbb{C}$.

Definition 1.1

A *linear vector space*, or simply a *vector space*, over $\mathbb{F}$ is a set X on which two operations, *addition*

$$+ : X \times X \to X,$$

and *scalar multiplication*

$$\cdot : \mathbb{F} \times X \to X,$$

are defined such that:

1. X is a commutative group under addition; that is,
   (a) $\mathbf{x}+\mathbf{y}=\mathbf{y}+\mathbf{x}$ for all $\mathbf{x},\mathbf{y} \in X$.
   (b) $\mathbf{x}+(\mathbf{y}+\mathbf{z})=(\mathbf{x}+\mathbf{y})+\mathbf{z}$ for all $\mathbf{x},\mathbf{y},\mathbf{z} \in X$.

(c) There is a *zero*, or *null*, *element* $\mathbf{0} \in X$ such that $\mathbf{x} + \mathbf{0} = \mathbf{x}$ for all $\mathbf{x} \in X$.

(d) For each $\mathbf{x} \in X$ there is an *additive inverse* $-\mathbf{x} \in X$ such that $\mathbf{x} + (-\mathbf{x}) = \mathbf{0}$.

2. Scalar multiplication between the elements of $\mathbb{F}$ and X satisfies

(a) $a \cdot (b \cdot \mathbf{x}) = (ab) \cdot \mathbf{x}$ for all $a, b \in \mathbb{F}$ and all $\mathbf{x} \in X$,

(b) $1 \cdot \mathbf{x} = \mathbf{x}$ for all $\mathbf{x} \in X$.

3. The two distributive properties

(a) $a \cdot (\mathbf{x} + \mathbf{y}) = a \cdot \mathbf{x} + a \cdot \mathbf{y}$

(b) $(a + b) \cdot \mathbf{x} = a \cdot \mathbf{x} + b \cdot \mathbf{x}$

hold for any $a, b \in \mathbb{F}$ and $\mathbf{x}, \mathbf{y} \in X$.

X is called a *real vector space* or a *complex vector space* depending on whether $\mathbb{F} = \mathbb{R}$ or $\mathbb{F} = \mathbb{C}$. The elements of X are called *vectors* and those of $\mathbb{F}$ *scalars.*

From these properties it can be shown that the zero vector $\mathbf{0}$ is unique, and that every $\mathbf{x} \in X$ has a unique inverse $-\mathbf{x}$. Furthermore, it follows that $0 \cdot \mathbf{x} = \mathbf{0}$ and $(-1) \cdot \mathbf{x} = -\mathbf{x}$ for every $\mathbf{x} \in X$, and that $a \cdot \mathbf{0} = \mathbf{0}$ for every $a \in \mathbb{F}$. As usual, we often drop the multiplication dot in $a \cdot \mathbf{x}$ and write $a\mathbf{x}$.

Example 1.2

(i) The set of n-tuples of real numbers

$$\mathbb{R}^n = \{(x_1, \ldots, x_n) : x_i \in \mathbb{R}\},$$

under addition, defined by

$$(x_1, \ldots, x_n) + (y_1, \ldots, y_n) = (x_1 + y_1, \ldots, x_n + y_n),$$

and scalar multiplication, defined by

$$a \cdot (x_1, \ldots, x_n) = (ax_1, \ldots, ax_n),$$

where $a \in \mathbb{R}$, is a real vector space.

(ii) The set of n-tuples of complex numbers

$$\mathbb{C}^n = \{(z_1, \ldots, z_n) : z_i \in \mathbb{C}\},$$

on the other hand, under the operations

$$\begin{aligned}(z_1, \dots, z_n) + (w_1, \dots, w_n) &= (z_1 + w_1, \dots, z_n + w_n), \\ a \cdot (z_1, \dots, z_n) &= (az_1, \dots, az_n), \quad a \in \mathbb{C},\end{aligned}$$

is a complex vector space.

(iii) The set $\mathbb{C}^n$ over the field $\mathbb{R}$ is a real vector space.

(iv) Let I be a real interval which may be closed, open, half-open, finite, or infinite. $\mathcal{P}(I)$ denotes the set of polynomials on I with real (complex) coefficients. This becomes a real (complex) vector space under the usual operation of addition of polynomials, and scalar multiplication

$$b \cdot (a_n x^n + \cdots + a_1 x + a_0) = ba_n x^n + \cdots + ba_1 x + ba_0,$$

where b is a real (complex) number. We also abbreviate $\mathcal{P}(\mathbb{R})$ as $\mathcal{P}$.

(v) The set of real (complex) continuous functions on the real interval I, which is denoted $C(I)$, is a real (complex) vector space under the usual operations of addition of functions and multiplication of a function by a real (complex) number.

Let $\{\mathbf{x}_1, \dots, \mathbf{x}_n\}$ be any finite set of vectors in a vector space X. The sum

$$a_1\mathbf{x}_1 + \cdots + a_n\mathbf{x}_n = \sum_{i=1}^{n} a_i\mathbf{x}_i, \quad a_i \in \mathbb{F},$$

is called a *linear combination* of the vectors in the set, and the scalars a_i are the *coefficients* in the linear combination.

Definition 1.3

(i) A finite set of vectors $\{\mathbf{x}_1, \dots, \mathbf{x}_n\}$ is said to be *linearly independent* if

$$\sum_{i=1}^{n} a_i\mathbf{x}_i = \mathbf{0} \Rightarrow a_i = 0 \quad \text{for all } i \in \{1, \dots, n\},$$

that is, if every linear combination of the vectors is not equal to zero except when all the coefficients are zeros. The set $\{\mathbf{x}_1, \dots, \mathbf{x}_n\}$ is *linearly dependent* if it is not linearly independent, that is, if there is a collection of coefficients $a_1, \dots, a_n$, not all zeros, such that $\sum_{i=1}^{n} a_i\mathbf{x}_i = \mathbf{0}$.

(ii) An infinite set of vectors $\{\mathbf{x}_1, \mathbf{x}_2, \mathbf{x}_3, \dots\}$ is *linearly independent* if every finite subset of the set is linearly independent. It is *linearly dependent* if it is not linearly independent, that is, if there is a finite subset of $\{\mathbf{x}_1, \mathbf{x}_2, \mathbf{x}_3, \dots\}$ which is linearly dependent.

It should be noted at this point that a finite set of vectors is linearly dependent if, and only if, one of the vectors can be represented as a linear combination of the others (see Exercise 1.3).

Definition 1.4

Let X be a vector space.

(i) A set $\mathcal{A}$ of vectors in X is said to *span* X if every vector in X can be expressed as a linear combination of elements of $\mathcal{A}$. If, in addition, the vectors in $\mathcal{A}$ are linearly independent, then $\mathcal{A}$ is called a *basis* of X.

(ii) A subset Y of X is called a *subspace* of X if every linear combination of vectors in Y lies in Y. This is equivalent to saying that Y is a vector space in its own right (over the same scalar field as X).

If X has a finite basis then any other basis of X is also finite, and both bases have the same number of elements (Exercise 1.4). This number is called the *dimension* of X and is denoted $\dim X$. If the basis is infinite, we take $\dim X = \infty$.

In Example 1.2, the vectors

$$\begin{aligned} \mathbf{e}_1 &= (1, 0, \ldots, 0), \\ \mathbf{e}_2 &= (0, 1, 0, \ldots, 0), \\ &\vdots \\ \mathbf{e}_n &= (0, \ldots, 0, 1) \end{aligned}$$

form a basis for $\mathbb{R}^n$ over $\mathbb{R}$ and $\mathbb{C}^n$ over $\mathbb{C}$. The vectors

$$\begin{aligned} \mathbf{d}_1 &= (i, 0, \ldots, 0), \\ \mathbf{d}_2 &= (0, i, 0, \ldots, 0), \\ &\vdots \\ \mathbf{d}_n &= (0, \ldots, 0, i), \end{aligned}$$

together with $\mathbf{e}_1, \ldots, \mathbf{e}_n$, form a basis of $\mathbb{C}^n$ over $\mathbb{R}$. On the other hand, the powers of $x \in \mathbb{R}$,

$$1,\ x,\ x^2,\ x^3, \ldots,$$

span $\mathcal{P}$ and, being linearly independent (Exercise 1.5), they form a basis for the space of real (complex) polynomials over $\mathbb{R}$ ($\mathbb{C}$). Thus both real $\mathbb{R}^n$ and complex $\mathbb{C}^n$ have dimension n, whereas real $\mathbb{C}^n$ has dimension $2n$. The space of polynomials, on the other hand, has infinite dimension. So does the space of continuous functions $C(I)$, as it includes all the polynomials on I (Exercise 1.6).

Let $\mathcal{P}_n(I)$ be the vector space of polynomials on the interval I of degree $\leq n$. This is clearly a subspace of $\mathcal{P}(I)$ of dimension $n+1$. Similarly, if we denote the set of (real or complex) functions on I whose first derivatives are continuous by $C^1(I)$, then, under the usual operations of addition of functions and multiplication by scalars, $C^1(I)$ is a vector subspace of $C(I)$ over the same (real or complex) field. As usual, when I is closed at one (or both) of its endpoints, the derivative at that endpoint is the one-sided derivative. More generally, by defining

$$C^n(I) = \{f \in C(I) : f^{(n)} \in C(I),\ n \in \mathbb{N}\},$$
$$C^\infty(I) = \bigcap_{n=1}^{\infty} C^n(I),$$

we obtain a sequence of vector spaces

$$C(I) \supset C^1(I) \supset C^2(I) \supset \cdots \supset C^\infty(I)$$

such that $C^k(I)$ is a (proper) vector subspace of $C^m(I)$ whenever $k > m$. Here $\mathbb{N}$ is the set of natural numbers $\{1, 2, 3, \ldots\}$ and $\mathbb{N}_0 = \mathbb{N} \cup \{0\}$. The set of integers $\{\ldots, -2, -1, 0, 1, 2, \ldots\}$ is denoted $\mathbb{Z}$. If we identify $C^0(I)$ with $C(I)$, all the spaces $C^n(I)$, $n \in \mathbb{N}_0$, have infinite dimensions as each includes the polynomials $\mathcal{P}(I)$. When $I = \mathbb{R}$, or when I is not relevant, we simply write C^n.

EXERCISES

1.1 Use the properties of the vector space X over $\mathbb{F}$ to prove the following.

(a) $0 \cdot \mathbf{x} = \mathbf{0}$ for all $\mathbf{x} \in X$.

(b) $a \cdot \mathbf{0} = \mathbf{0}$ for all $a \in \mathbb{F}$.

(c) $(-1) \cdot \mathbf{x} = -\mathbf{x}$ for all $\mathbf{x} \in X$.

(d) If $a \cdot \mathbf{x} = \mathbf{0}$ then either $a = 0$ or $\mathbf{x} = \mathbf{0}$.

1.2 Determine which of the following sets is a vector space under the usual operations of addition and scalar multiplication, and whether it is a real or a complex vector space.

(a) $\mathcal{P}_n(I)$ with complex coefficients over $\mathbb{C}$

(b) $\mathcal{P}(I)$ with imaginary coefficients over $\mathbb{R}$

(c) The set of real numbers over $\mathbb{C}$

(d) The set of complex functions of class $C^n(I)$ over $\mathbb{R}$

1.3 Prove that the vectors $\mathbf{x}_1, \ldots, \mathbf{x}_n$ are linearly dependent if, and only if, there is an integer $k \in \{1, \ldots, n\}$ such that

$$\mathbf{x}_k = \sum_{i \neq k} a_i \mathbf{x}_i, \qquad a_i \in \mathbb{F}.$$

Conclude from this that any set of vectors, whether finite or infinite, is linearly dependent if, and only if, one of its vectors is a finite linear combination of the other vectors.

1.4 Let X be a vector space. Prove that, if $\mathcal{A}$ and $\mathcal{B}$ are bases of X and one of them is finite, then so is the other and they have the same number of elements.

1.5 Show that any finite set of powers of x, $\{1, x, x^2, \ldots, x^n : x \in I\}$, is linearly independent. Hence conclude that the infinite set $\{1, x, x^2, \ldots : x \in I\}$ is linearly independent.

1.6 If Y is a subspace of the vector space X, prove that $\dim Y \leq \dim X$.

1.7 Prove that the vectors

$$\mathbf{x}_1 = (x_{11}, \ldots, x_{1n}),$$
$$\vdots$$
$$\mathbf{x}_n = (x_{n1}, \ldots, x_{nn}),$$

where $x_{ij} \in \mathbb{R}$, are linearly dependent if, and only if, $\det(x_{ij}) = 0$, where $\det(x_{ij})$ is the determinant of the matrix (x_{ij}).

1.2 Inner Product Space

Definition 1.5

Let X be a vector space over $\mathbb{F}$. A function from $X \times X$ to $\mathbb{F}$ is called an *inner product* in X if, for any pair of vectors $\mathbf{x}, \mathbf{y} \in X$, the inner product $(\mathbf{x}, \mathbf{y}) \mapsto \langle \mathbf{x}, \mathbf{y} \rangle \in \mathbb{F}$ satisfies the following conditions.

(i) $\langle \mathbf{x}, \mathbf{y} \rangle = \overline{\langle \mathbf{y}, \mathbf{x} \rangle}$ for all $\mathbf{x}, \mathbf{y} \in X$.

(ii) $\langle a\mathbf{x}+b\mathbf{y}, \mathbf{z} \rangle = a \langle \mathbf{x}, \mathbf{z} \rangle + b \langle \mathbf{y}, \mathbf{z} \rangle$ for all $a, b \in \mathbb{F}$, $\mathbf{x}, \mathbf{y}, \mathbf{z} \in X$.

(iii) $\langle \mathbf{x}, \mathbf{x} \rangle \geq 0$ for all $\mathbf{x} \in X$.

(iv) $\langle \mathbf{x}, \mathbf{x} \rangle = 0 \Leftrightarrow \mathbf{x} = \mathbf{0}$.

A vector space on which an inner product is defined is called an *inner product space*.

The symbol $\overline{\langle \mathbf{y}, \mathbf{x} \rangle}$ in (i) denotes the complex conjugate of $\langle \mathbf{y}, \mathbf{x} \rangle$, so that $\langle \mathbf{x}, \mathbf{y} \rangle = \langle \mathbf{y}, \mathbf{x} \rangle$ if X is a real vector space. Note also that (i) and (ii) imply

$$\langle \mathbf{x}, a\mathbf{y} \rangle = \overline{\langle a\mathbf{y}, \mathbf{x} \rangle} = \bar{a} \langle \mathbf{x}, \mathbf{y} \rangle,$$

which means that the linearity property which holds in the first component of the inner product, as expressed by (ii), does not apply to the second component unless $\mathbb{F} = \mathbb{R}$.

Theorem 1.6 (Cauchy–Bunyakowsky–Schwarz Inequality)

If X is an inner product space, then

$$|\langle \mathbf{x}, \mathbf{y} \rangle|^2 \leq \langle \mathbf{x}, \mathbf{x} \rangle \langle \mathbf{y}, \mathbf{y} \rangle \quad \text{for all } \mathbf{x}, \mathbf{y} \in X.$$

Proof

If either $\mathbf{x} = \mathbf{0}$ or $\mathbf{y} = \mathbf{0}$ this inequality clearly holds, so we need only consider the case where $\mathbf{x} \neq \mathbf{0}$ and $\mathbf{y} \neq \mathbf{0}$. Furthermore, neither side of the inequality is affected if we replace $\mathbf{x}$ by $a\mathbf{x}$ where $|a| = 1$. Choose a so that $\langle a\mathbf{x}, \mathbf{y} \rangle$ is a real number; that is, if $\langle \mathbf{x}, \mathbf{y} \rangle = |\langle \mathbf{x}, \mathbf{y} \rangle| e^{i\theta}$, let $a = e^{-i\theta}$. Therefore we may assume, without loss of generality, that $\langle \mathbf{x}, \mathbf{y} \rangle$ is a real number. Using the above properties of the inner product, we have, for any real number t,

$$0 \leq \langle \mathbf{x}+t\mathbf{y}, \mathbf{x}+t\mathbf{y} \rangle = \langle \mathbf{x}, \mathbf{x} \rangle + 2 \langle \mathbf{x}, \mathbf{y} \rangle t + \langle \mathbf{y}, \mathbf{y} \rangle t^2. \tag{1.1}$$

This is a real quadratic expression in t which achieves its minimum at $t = -\langle \mathbf{x}, \mathbf{y} \rangle / \langle \mathbf{y}, \mathbf{y} \rangle$. Substituting this value for t into (1.1) gives

$$0 \leq \langle \mathbf{x}, \mathbf{x} \rangle - \frac{\langle \mathbf{x}, \mathbf{y} \rangle^2}{\langle \mathbf{y}, \mathbf{y} \rangle},$$

and hence the desired inequality. □

We now define the *norm* of the vector $\mathbf{x}$ as

$$\|\mathbf{x}\| = \sqrt{\langle \mathbf{x}, \mathbf{x} \rangle}.$$

Hence, in view of (iii) and (iv), $\|\mathbf{x}\| \geq 0$ for all $\mathbf{x} \in X$, and $\|\mathbf{x}\| = 0$ if and only if $\mathbf{x} = \mathbf{0}$. The Cauchy–Bunyakowsky–Schwarz inequality, which we henceforth refer to as the CBS inequality, then takes the form

$$|\langle \mathbf{x}, \mathbf{y} \rangle| \leq \|\mathbf{x}\| \, \|\mathbf{y}\| \quad \text{for all } \mathbf{x}, \mathbf{y} \in X. \tag{1.2}$$

Corollary 1.7

If X is an inner product space, then

$$\|\mathbf{x}+\mathbf{y}\| \leq \|\mathbf{x}\| + \|\mathbf{y}\| \quad \text{for all } \mathbf{x},\mathbf{y} \in X. \tag{1.3}$$

Proof

By definition of the norm,

$$\begin{aligned}
\|\mathbf{x}+\mathbf{y}\|^2 &= \langle \mathbf{x}+\mathbf{y}, \mathbf{x}+\mathbf{y} \rangle \\
&= \|\mathbf{x}\|^2 + \langle \mathbf{x},\mathbf{y} \rangle + \langle \mathbf{y},\mathbf{x} \rangle + \|\mathbf{y}\|^2 \\
&= \|\mathbf{x}\|^2 + 2\,\mathrm{Re}\,\langle \mathbf{x},\mathbf{y} \rangle + \|\mathbf{y}\|^2 .
\end{aligned}$$

But $\mathrm{Re}\,\langle \mathbf{x},\mathbf{y} \rangle \leq |\langle \mathbf{x},\mathbf{y} \rangle| \leq \|\mathbf{x}\|\,\|\mathbf{y}\|$ by the CBS inequality, hence

$$\begin{aligned}
\|\mathbf{x}+\mathbf{y}\|^2 &\leq \|\mathbf{x}\|^2 + 2\,\|\mathbf{x}\|\,\|\mathbf{y}\| + \|\mathbf{y}\|^2 \\
&= (\|\mathbf{x}\| + \|\mathbf{y}\|)^2.
\end{aligned}$$

Inequality (1.3) now follows by taking the square roots of both sides. □

By defining the *distance* between the vectors $\mathbf{x}$ and $\mathbf{y}$ to be $\|\mathbf{x}-\mathbf{y}\|$, we see that for any three vectors $\mathbf{x},\mathbf{y},\mathbf{z} \in X$,

$$\begin{aligned}
\|\mathbf{x}-\mathbf{y}\| &= \|\mathbf{x}-\mathbf{z}+\mathbf{z}-\mathbf{y}\| \\
&\leq \|\mathbf{x}-\mathbf{z}\| + \|\mathbf{z}-\mathbf{y}\| .
\end{aligned}$$

This inequality, and by extension (1.3), is called the *triangle inequality,* as it generalizes a well known inequality between the sides of a triangle in the plane whose vertices are the points $\mathbf{x},\mathbf{y},\mathbf{z}$. The inner product space X is now a *topological space*, in which the topology is defined by the norm $\|\cdot\|$, which is derived from the inner product $\langle \cdot,\cdot \rangle$.

Example 1.8

(a) In $\mathbb{R}^n$ we define the inner product of the vectors

$$\mathbf{x} = (x_1, \ldots, x_n), \quad \mathbf{y} = (y_1, \ldots, y_n)$$

by

$$\langle \mathbf{x},\mathbf{y} \rangle = x_1 y_1 + \ldots + x_n y_n, \tag{1.4}$$

which implies

$$\|\mathbf{x}\| = \sqrt{x_1^2 + \cdots + x_n^2}.$$

In this topology the vector space $\mathbb{R}^n$ is the familiar n-dimensional Euclidean space. Note that there are other choices for defining the inner product $\langle \mathbf{x}, \mathbf{y} \rangle$, such as $c(x_1y_1 + \cdots + x_ny_n)$ where c is any positive number, or $c_1x_1y_1 + \cdots + c_nx_ny_n$ where $c_i > 0$ for every i. In either case the provisions of Definition 1.5 are all satisfied, but the resulting inner product space would not in general be Euclidean.

(b) In $\mathbb{C}^n$ we define

$$\langle \mathbf{z}, \mathbf{w} \rangle = z_1\bar{w}_1 + \cdots + z_n\bar{w}_n \tag{1.5}$$

for any pair $\mathbf{z}, \mathbf{w} \in \mathbb{C}^n$. Consequently,

$$\|\mathbf{z}\| = \sqrt{|z_1|^2 + \cdots + |z_n|^2}.$$

(c) A natural choice for the definition of an inner product on $C([a,b])$, by analogy with (1.5), is

$$\langle f, g \rangle = \int_a^b f(x)\overline{g(x)}dx, \qquad f, g \in C([a,b]), \tag{1.6}$$

so that

$$\|f\| = \left[\int_a^b |f(x)|^2 \, dx\right]^{1/2}.$$

It is a simple matter to verify that the properties (i) through (iv) of the inner product are satisfied in each case, provided of course that $\mathbb{F}= \mathbb{C}$ when the vector space is $\mathbb{C}^n$ or complex $C([a,b])$. To check (iv) in Example 1.8(c), we have to show that

$$\left[\int_a^b |f(x)|^2 \, dx\right]^{1/2} = 0 \quad \Leftrightarrow \quad f(x) = 0 \quad \text{for all } x \in [a,b].$$

We need only verify the forward implication ($\Rightarrow$), as the backward implication ($\Leftarrow$) is trivial. But this follows from a well-known property of continuous, nonnegative functions: If φ is continuous on $[a,b]$, $\varphi \geq 0$, and $\int_a^b \varphi(x)dx = 0$, then $\varphi = 0$ (see [1], for example). Because $|f|^2$ is continuous and nonnegative on $[a,b]$ for any $f \in C([a,b])$,

$$\|f\| = 0 \Rightarrow \int_a^b |f(x)|^2 \, dx = 0 \Rightarrow |f|^2 = 0 \Rightarrow f = 0.$$

In this study, we are mainly concerned with function spaces on which an inner product of the type (1.6) is defined. In addition to the topological structure which derives from the norm $\|\cdot\|$, this inner product endows the space with a geometrical structure that extends some desirable notions, such as orthogonality, from Euclidean space to infinite-dimensional spaces. This is taken up in Section 1.3. Here we examine the Euclidean space $\mathbb{R}^n$ more closely.

Although we proved the CBS and the triangle inequalities for any inner product in Theorem 1.6 and its corollary, we can also derive these inequalities directly in $\mathbb{R}^n$. Consider the inequality

$$(a-b)^2 = a^2 - 2ab + b^2 \geq 0 \tag{1.7}$$

which holds for any pair of real numbers a and b. Let

$$a = \frac{x_i}{\sqrt{x_1^2 + \cdots + x_n^2}}, \quad b = \frac{y_i}{\sqrt{y_1^2 + \cdots + y_n^2}}, \quad x_i, y_i \in \mathbb{R}.$$

If $\sum_{j=1}^n x_j^2 \neq 0$ and $\sum_{j=1}^n y_j^2 \neq 0$, then (1.7) implies

$$\frac{x_i y_i}{\sqrt{\sum x_j^2}\sqrt{\sum y_j^2}} \leq \frac{1}{2}\frac{x_i^2}{\sum x_j^2} + \frac{1}{2}\frac{y_i^2}{\sum y_j^2},$$

where the summation over the index j is from 1 to n. After summing on i from 1 to n, the right-hand side of this inequality reduces to 1, and we obtain

$$\sum x_i y_i \leq \sqrt{\sum x_i^2}\sqrt{\sum y_i^2}.$$

This inequality remains valid regardless of the signs of x_i and y_i, therefore we can write

$$\left|\sum x_i y_i\right| \leq \sqrt{\sum x_i^2}\sqrt{\sum y_i^2}$$

for all $\mathbf{x} = (x_1, \ldots, x_n) \neq \mathbf{0}$ and $\mathbf{y} = (y_1, \ldots, y_n) \neq \mathbf{0}$ in $\mathbb{R}^n$. But because the inequality becomes an equality if either $\|\mathbf{x}\|$ or $\|\mathbf{y}\|$ is 0, this proves the CBS inequality

$$|\langle \mathbf{x}, \mathbf{y}\rangle| \leq \|\mathbf{x}\| \, \|\mathbf{y}\| \quad \text{for all } \mathbf{x}, \mathbf{y} \in \mathbb{R}^n.$$

From this the triangle inequality $\|\mathbf{x}+\mathbf{y}\| \leq \|\mathbf{x}\| + \|\mathbf{y}\|$ immediately follows.

Now we define the angle $\theta \in [0, \pi]$ between any pair of nonzero vectors $\mathbf{x}$ and $\mathbf{y}$ in $\mathbb{R}^n$ by the equation

$$\langle \mathbf{x}, \mathbf{y}\rangle = \|\mathbf{x}\| \, \|\mathbf{y}\| \cos\theta.$$

Because the function $\cos : [0, \pi] \to [-1, 1]$ is injective, this defines the angle θ uniquely and agrees with the usual definition of the angle between $\mathbf{x}$ and $\mathbf{y}$ in both $\mathbb{R}^2$ and $\mathbb{R}^3$. With $\mathbf{x} \neq \mathbf{0}$ and $\mathbf{y} \neq \mathbf{0}$,

$$\langle \mathbf{x}, \mathbf{y}\rangle = 0 \quad \Leftrightarrow \quad \cos\theta = 0,$$

which is the condition for the vectors $\mathbf{x}, \mathbf{y} \in \mathbb{R}^n$ to be orthogonal. Consequently, we adopt the following definition.

Definition 1.9

(i) A pair of nonzero vectors $\mathbf{x}$ and $\mathbf{y}$ in the inner product space X is said to be *orthogonal* if $\langle \mathbf{x}, \mathbf{y} \rangle = 0$, symbolically expressed by writing $\mathbf{x} \perp \mathbf{y}$. A set of nonzero vectors $\mathcal{V}$ in X is *orthogonal* if every pair in $\mathcal{V}$ is orthogonal.

(ii) An orthogonal set $\mathcal{V} \subseteq X$ is said to be *orthonormal* if $\|\mathbf{x}\| = 1$ for every $\mathbf{x} \in \mathcal{V}$.

A typical example of an orthonormal set in the Euclidean space $\mathbb{R}^n$ is given by

$$\begin{aligned} \mathbf{e}_1 &= (1, 0, \ldots, 0), \\ \mathbf{e}_2 &= (0, 1, \ldots, 0), \\ &\vdots \\ \mathbf{e}_n &= (0, \ldots, 0, 1), \end{aligned}$$

which, as we have already seen, forms a basis of $\mathbb{R}^n$.

In general, if the vectors

$$\mathbf{x}_1, \mathbf{x}_2, \ldots, \mathbf{x}_n \tag{1.8}$$

in the inner product space X are orthogonal, then they are necessarily linearly independent. To see that, let

$$a_1 \mathbf{x}_1 + \cdots + a_n \mathbf{x}_n = \mathbf{0}, \quad a_i \in \mathbb{F},$$

and take the inner product of each side of this equation with $\mathbf{x}_k$, $1 \le k \le n$. In as much as $\langle \mathbf{x}_i, \mathbf{x}_k \rangle = 0$ whenever $i \neq k$, we obtain

$$\begin{aligned} a_k \langle \mathbf{x}_k, \mathbf{x}_k \rangle = a_k \|\mathbf{x}_k\|^2 = 0, \quad k \in \{1, \cdots, n\} \\ \Rightarrow a_k = 0 \quad \text{for all } k. \end{aligned}$$

By dividing each vector in (1.8) by its norm, we obtain the orthonormal set $\{\mathbf{x}_i / \|\mathbf{x}_i\| : 1 \le i \le n\}$.

Let us go back to the Euclidean space $\mathbb{R}^n$ and assume that $\mathbf{x}$ is any vector in $\mathbb{R}^n$. We can therefore represent it in the basis $\{\mathbf{e}_1, \ldots, \mathbf{e}_n\}$ by

$$\mathbf{x} = \sum_{i=1}^{n} a_i \mathbf{e}_i. \tag{1.9}$$

Taking the inner product of Equation (1.9) with $\mathbf{e}_k$, and using the orthonormal property of $\{\mathbf{e}_i\}$,

$$\langle \mathbf{x}, \mathbf{e}_k \rangle = a_k, \quad k \in \{1, \ldots, n\}.$$

This determines the coefficients a_i in (1.9), and means that any vector $\mathbf{x}$ in $\mathbb{R}^n$ is represented by the formula

$$\mathbf{x} = \sum_{i=1}^{n} \langle \mathbf{x}, \mathbf{e}_i \rangle \mathbf{e}_i.$$

The number $\langle \mathbf{x}, \mathbf{e}_i \rangle$ is called the *projection* of $\mathbf{x}$ on $\mathbf{e}_i$, and $\langle \mathbf{x}, \mathbf{e}_i \rangle \mathbf{e}_i$ is the *projection vector* in the direction of $\mathbf{e}_i$. More generally, if $\mathbf{x}$ and $\mathbf{y} \neq \mathbf{0}$ are any vectors in the inner product space X, then $\langle \mathbf{x}, \mathbf{y}/\|\mathbf{y}\| \rangle$ is the projection of $\mathbf{x}$ on $\mathbf{y}$, and the vector

$$\left\langle \mathbf{x}, \frac{\mathbf{y}}{\|\mathbf{y}\|} \right\rangle \frac{\mathbf{y}}{\|\mathbf{y}\|} = \frac{\langle \mathbf{x}, \mathbf{y} \rangle}{\|\mathbf{y}\|^2} \mathbf{y}$$

is its projection vector along $\mathbf{y}$.

Suppose now that we have a linearly independent set of vectors $\{\mathbf{x}_1, \ldots, \mathbf{x}_n\}$ in the inner product space X. Can we form an orthogonal set out of this set? In what follows we present the so-called Gram–Schmidt method for constructing an orthogonal set $\{\mathbf{y}_1, \ldots, \mathbf{y}_n\}$ out of $\{\mathbf{x}_i\}$ having the same number of vectors: We first choose

$$\mathbf{y}_1 = \mathbf{x}_1.$$

The second vector is obtained from $\mathbf{x}_2$ after extracting the projection vector of $\mathbf{x}_2$ in the direction of $\mathbf{y}_1$,

$$\mathbf{y}_2 = \mathbf{x}_2 - \frac{\langle \mathbf{x}_2, \mathbf{y}_1 \rangle}{\|\mathbf{y}_1\|^2} \mathbf{y}_1.$$

The third vector is $\mathbf{x}_3$ minus the projections of $\mathbf{x}_3$ in the directions of $\mathbf{y}_1$ and $\mathbf{y}_2$,

$$\mathbf{y}_3 = \mathbf{x}_3 - \frac{\langle \mathbf{x}_3, \mathbf{y}_1 \rangle}{\|\mathbf{y}_1\|^2} \mathbf{y}_1 - \frac{\langle \mathbf{x}_3, \mathbf{y}_2 \rangle}{\|\mathbf{y}_2\|^2} \mathbf{y}_2.$$

We continue in this fashion until the last vector

$$\mathbf{y}_n = \mathbf{x}_n - \frac{\langle \mathbf{x}_n, \mathbf{y}_1 \rangle}{\|\mathbf{y}_1\|^2} \mathbf{y}_1 - \cdots - \frac{\langle \mathbf{x}_n, \mathbf{y}_{n-1} \rangle}{\|\mathbf{y}_{n-1}\|^2} \mathbf{y}_{n-1},$$

and the reader can verify that the set $\{\mathbf{y}_1, \ldots, \mathbf{y}_n\}$ is orthogonal.

EXERCISES

1.8 Given two vectors $\mathbf{x}$ and $\mathbf{y}$ in an inner product space, under what conditions does the equality $\|\mathbf{x} + \mathbf{y}\|^2 = \|\mathbf{x}\|^2 + \|\mathbf{y}\|^2$ hold? Can this equation hold even if the vectors are not orthogonal?

1.9 Let $\mathbf{x}, \mathbf{y} \in X$, where X is an inner product space.

(a) If the vectors $\mathbf{x}$ and $\mathbf{y}$ are linearly independent, prove that $\mathbf{x}+\mathbf{y}$ and $\mathbf{x}-\mathbf{y}$ are also linearly independent.

(b) If $\mathbf{x}$ and $\mathbf{y}$ are orthogonal and nonzero, when are $\mathbf{x}+\mathbf{y}$ and $\mathbf{x}-\mathbf{y}$ orthogonal?

1.10 Let $\varphi_1(x) = 1,\ \varphi_2(x) = x,\ \varphi_3(x) = x^2,\ -1 \leq x \leq 1$. Use (1.6) to calculate

(a) $\langle \varphi_1, \varphi_2 \rangle$

(b) $\langle \varphi_1, \varphi_3 \rangle$

(c) $\|\varphi_1 - \varphi_2\|^2$

(d) $\|2\varphi_1 + 3\varphi_2\|$.

1.11 Determine all orthogonal pairs on $[0,1]$ among the functions $\varphi_1(x) = 1,\ \varphi_2(x) = x,\ \varphi_3(x) = \sin 2\pi x,\ \varphi_4(x) = \cos 2\pi x$. What is the largest orthogonal subset of $\{\varphi_1, \varphi_2, \varphi_3, \varphi_4\}$?

1.12 Determine the projection of $f(x) = \cos^2 x$ on each of the functions $f_1(x) = 1,\ f_2(x) = \cos x,\ f_3(x) = \cos 2x,\ -\pi \leq x \leq \pi$.

1.13 Verify that the functions $\varphi_1, \varphi_2, \varphi_3$ in Exercise 1.10 are linearly independent, and use the Gram–Schmidt method to construct a corresponding orthogonal set.

1.14 Prove that the set of functions $\{1, x, |x|\}$ is linearly independent on $[-1,1]$, and construct a corresponding orthonormal set. Is the given set linearly independent on $[0,1]$?

1.15 Use the result of Exercise 1.3 and the properties of determinants to prove that any set of functions $\{f_1, \ldots, f_n\}$ in $C^{n-1}(I)$, I being a real interval, is linearly dependent if, and only if, $\det(f_i^{(j)}) = 0$ on I, where $1 \leq i \leq n, 0 \leq j \leq n-1$.

1.16 Verify that the following functions are orthogonal on $[-1,1]$.

$$\varphi_1(x) = 1, \quad \varphi_2(x) = x^2 - \frac{1}{3}, \quad \varphi_3(x) = \begin{cases} x/|x|, & x \neq 0 \\ 0, & x = 0. \end{cases}$$

Determine the corresponding orthonormal set.

1.17 Determine the values of the coefficients a and b which make the function $x^2 + ax + b$ orthogonal to both $x+1$ and $x-1$ on $[0,1]$.

1.18 Using the definition of the inner product as expressed by Equation (1.6), show that $\|f\| = 0$ does not necessarily imply that $f = 0$ unless f is continuous.

1.3 The Space $\mathcal{L}^2$

For any two functions f and g in the vector space $C([a,b])$ of complex continuous functions on a real interval $[a,b]$, we defined the inner product

$$\langle f,g\rangle = \int_a^b f(x)\overline{g(x)}dx, \tag{1.10}$$

from which followed the definition of the norm

$$\|f\| = \sqrt{\langle f,f\rangle} = \sqrt{\int_a^b |f(x)|^2\, dx}. \tag{1.11}$$

As in $\mathbb{R}^n$, we can also show directly that the CBS inequality holds in $C([a,b])$. For any $f,g \in C([a,b])$, we have

$$\left\| \frac{|f|}{\|f\|} - \frac{|g|}{\|g\|} \right\|^2 = \int_a^b \left[\frac{|f(x)|}{\|f\|} - \frac{|g(x)|}{\|g\|} \right]^2 dx \geq 0,$$

where we assume that $\|f\| \neq 0$ and $\|g\| \neq 0$. Hence

$$\int_a^b \frac{|f(x)|}{\|f\|}\frac{|g(x)|}{\|g\|}dx \leq \frac{1}{2\|f\|^2}\int_a^b |f(x)|^2\, dx + \frac{1}{2\|g\|^2}\int_a^b |g(x)|^2\, dx = 1$$

$$\Rightarrow \langle |f|, |g|\rangle \leq \|f\|\,\|g\|.$$

Using the monotonicity property of the integral

$$\left| \int_a^b \varphi(x)dx \right| \leq \int_a^b |\varphi(x)|\, dx,$$

we therefore conclude that

$$|\langle f,g\rangle| \leq \langle |f|, |g|\rangle \leq \|f\|\,\|g\|.$$

If either $\|f\| = 0$ or $\|g\| = 0$ the inequality remains valid, as it becomes an equality. The triangle inequality

$$\|f+g\| \leq \|f\| + \|g\|$$

then easily follows from the relation $f\bar{g} + \bar{f}g = 2\operatorname{Re} f\bar{g} \leq 2|fg|$.

As we have already observed, the nonnegative number $\|f-g\|$ may be regarded as a measure of the "distance" between the functions $f,g \in C([a,b])$. In this case we clearly have $\|f-g\| = 0$ if, and only if, $f = g$ on $[a,b]$. This is the advantage of dealing with continuous functions, for if we admit discontinuous functions, such as

$$h(x) = \begin{cases} 1, & x = 1 \\ 0, & x \in (1,2], \end{cases} \tag{1.12}$$

then $\|h\| = 0$ whereas $h \neq 0$.

Nevertheless, $C([a,b])$ is not a suitable inner product space for pursuing this study, for it is not closed under limit operations as we show in the next section. That is to say, if a sequence of functions in $C([a,b])$ "converges" (in a sense to be defined in Section 1.4) its "limit" may not be in $C([a,b])$. So we need to enlarge the space of continuous functions over $[a,b]$ in order to avoid this difficulty. But in this larger space, call it $X([a,b])$, we can only admit functions for which the inner product

$$\langle f,g\rangle = \int_a^b f(x)\overline{g(x)}dx$$

is defined for every pair $f,g \in X([a,b])$. Now the CBS inequality $|\langle f,g\rangle| \leq \|f\|\,\|g\|$ ensures that the inner product of f and g is well defined if $\|f\|$ and $\|g\|$ exist (i.e., if $|f|^2$ and $|g|^2$ are integrable). Strictly speaking, this is only true if the integrals are interpreted as Lebesgue integrals, for the Riemann integrability of f^2 and g^2 does not guarantee the Riemann integrability of fg (see Exercise 1.21); but in this study we shall have no occasion to deal with functions which are integrable in the sense of Lebesgue but not in the sense of Riemann. For our purposes, Riemann integration, and its extension to improper integrals, is adequate. The space $X([a,b])$ which we seek should therefore be made up of functions f such that $|f|^2$ is integrable on $[a,b]$.

We use the symbol $\mathcal{L}^2(a,b)$ to denote the set of functions $f:[a,b]\to\mathbb{C}$ such that

$$\int_a^b |f(x)|^2\,dx < \infty.$$

By defining the inner product (1.10) and the norm (1.11) on $\mathcal{L}^2(a,b)$, we can use the triangle inequality to obtain

$$\begin{aligned}\|\alpha f+\beta g\| &\leq \|\alpha f\| + \|\beta g\|\\ &= |\alpha|\,\|f\| + |\beta|\,\|g\| \quad \text{for all } f,g\in\mathcal{L}^2(a,b),\quad \alpha,\beta\in\mathbb{C},\end{aligned}$$

hence $\alpha f+\beta g\in\mathcal{L}^2(a,b)$ whenever $f,g\in\mathcal{L}^2(a,b)$. Thus $\mathcal{L}^2(a,b)$ is a linear vector space which, under the inner product (1.10), becomes an inner product space and includes $C([a,b])$ as a proper subspace.

In $\mathcal{L}^2(a,b)$ the equality $\|f\|=0$ does not necessarily mean $f(x)=0$ at every point $x\in[a,b]$. For example, in the case where $f(x)=0$ on all but a finite number of points in $[a,b]$ we clearly have $\|f\|=0$. We say that $f=0$ *pointwise* on a real interval I if $f(x)=0$ at every $x\in I$. If $\|f\|=0$ we say that $f=0$ *in* $\mathcal{L}^2(I)$. Thus the function h defined in (1.12) equals 0 in $\mathcal{L}^2(I)$, but not pointwise. The function 0 in $\mathcal{L}^2(I)$ really denotes an *equivalence class of functions*, each of which has norm 0. The function which is pointwise equal to 0 is only one member, indeed the only continuous member, of that class. Similarly, we say that two functions f and g in $\mathcal{L}^2(I)$ are *equal in* $\mathcal{L}^2(I)$ if $\|f-g\|=0$, although f and g may not be equal pointwise on I. In the terminology of measure theory, f and g are said to be "equal almost everywhere." Hence the space $\mathcal{L}^2(a,b)$ is, in fact, made up of equivalence classes of functions defined by equality in $\mathcal{L}^2(a,b)$, that is, functions which are equal almost everywhere.

Thus far we have used the symbol $\mathcal{L}^2(a,b)$ to denote the linear space of functions $f:[a,b]\to\mathbb{C}$ such that $\int_a^b |f(x)|^2\,dx<\infty$. But because this integral is not affected by replacing the closed interval $[a,b]$ by $[a,b)$, $(a,b]$, or (a,b), $\mathcal{L}^2(a,b)$ coincides with $\mathcal{L}^2([a,b))$, $\mathcal{L}^2((a,b])$ and $\mathcal{L}^2((a,b))$. The interval (a,b) need not be bounded at either or both ends, and so we have $\mathcal{L}^2(a,\infty)$, $\mathcal{L}^2(-\infty,b)$ and $\mathcal{L}^2(-\infty,\infty)=\mathcal{L}^2(\mathbb{R})$. In such cases, as in the case when the function is unbounded, we interpret the integral of $|f|^2$ on (a,b) as an improper Riemann integral. Sometimes we simply write $\mathcal{L}^2$ when the underlying interval is not specified or irrelevant to the discussion.

Example 1.10

Determine each function which belongs to $\mathcal{L}^2$ and calculate its norm.

(i) $f(x)=\begin{cases} 1, & 0\le x<1/2 \\ 0, & 1/2\le x\le 1.\end{cases}$

(ii) $f(x)=1/\sqrt{x},\ 0<x<1.$

(iii) $f(x)=1/\sqrt[3]{x},\ 0<x<1.$

(iv) $f(x)=1/x,\ 1<x<\infty.$

Solution

(i)

$$\|f\|^2=\int_0^1 f^2(x)dx=\int_0^{1/2}dx=\frac{1}{2}.$$

Therefore $f\in\mathcal{L}^2(0,1)$ and $\|f\|=1/\sqrt{2}$.

(ii)

$$\|f\|^2=\int_0^1\frac{1}{x}dx=\lim_{\varepsilon\to 0^+}\int_\varepsilon^1\frac{1}{x}dx=-\lim_{\varepsilon\to 0^+}\log\varepsilon=\infty$$
$$\Rightarrow f\notin\mathcal{L}^2(0,1).$$

(iii)

$$\|f\|^2=\int_0^1\frac{1}{x^{2/3}}dx=\lim_{\varepsilon\to 0^+}3(1-\varepsilon^{1/3})=3$$
$$\Rightarrow f\in\mathcal{L}^2(0,1),\ \|f\|=\sqrt{3}.$$

(iv)

$$\begin{aligned}\|f\|^2 &= \int_1^\infty \frac{1}{x^2}dx = \lim_{b\to\infty} -\left(\frac{1}{b}-1\right) = 1\\ &\Rightarrow f \in \mathcal{L}^2(1,\infty),\ \|f\| = 1.\end{aligned}$$

Example 1.11

The infinite set of functions $\{1, \cos x, \sin x, \cos 2x, \sin 2x, \ldots\}$ is orthogonal in the real inner product space $\mathcal{L}^2(-\pi,\pi)$. This can be seen by calculating the inner product of each pair in the set:

$$\begin{aligned}\langle 1, \cos nx\rangle &= \int_{-\pi}^{\pi} \cos nx\, dx = 0, \quad n \in \mathbb{N}.\\ \langle 1, \sin nx\rangle &= \int_{-\pi}^{\pi} \sin nx\, dx = 0, \quad n \in \mathbb{N}.\end{aligned}$$

$$\begin{aligned}\langle \cos nx, \cos mx\rangle &= \int_{-\pi}^{\pi} \cos nx \cos mx\, dx\\ &= \frac{1}{2}\int_{-\pi}^{\pi} [\cos(n-m)x + \cos(n+m)x]dx\\ &= \frac{1}{2}\left[\frac{1}{n-m}\sin(n-m)x + \frac{1}{n+m}\sin(n+m)x\right]\Bigg|_{-\pi}^{\pi}\\ &= 0, \quad n \neq m.\end{aligned}$$

$$\begin{aligned}\langle \sin nx, \sin mx\rangle &= \int_{-\pi}^{\pi} \sin nx \sin mx\, dx\\ &= \frac{1}{2}\int_{-\pi}^{\pi} [\cos(n-m)x - \cos(n+m)x]dx\\ &= 0, \quad n \neq m.\end{aligned}$$

$$\langle \cos nx, \sin mx\rangle = \int_{-\pi}^{\pi} \cos nx \sin mx\, dx = 0, \quad n, m \in \mathbb{N},$$

because $\cos nx \sin mx$ is an odd function. Furthermore,

$$\begin{aligned}\|1\| &= \sqrt{2\pi},\\ \|\cos nx\| &= \left[\int_{-\pi}^{\pi} \cos^2 nx\, dx\right]^{1/2} = \sqrt{\pi},\\ \|\sin nx\| &= \left[\int_{-\pi}^{\pi} \sin^2 nx\, dx\right]^{1/2} = \sqrt{\pi}, \quad n \in \mathbb{N}.\end{aligned}$$

Thus the set
$$\left\{\frac{1}{\sqrt{2\pi}}, \frac{\cos x}{\sqrt{\pi}}, \frac{\sin x}{\sqrt{\pi}}, \frac{\cos 2x}{\sqrt{\pi}}, \frac{\sin 2x}{\sqrt{\pi}}, \cdots\right\},$$
which is obtained by dividing each function in the orthogonal set by its norm, is orthonormal in $\mathcal{L}^2(-\pi,\pi)$.

Example 1.12

The set of functions
$$\{e^{inx} : n \in \mathbb{Z}\} = \{\ldots, e^{-i2x}, e^{-ix}, 1, e^{ix}, e^{i2x}, \ldots\}$$
is orthogonal in the complex space $\mathcal{L}^2(-\pi,\pi)$, because, for any $n \neq m$,
$$\begin{aligned}\left\langle e^{inx}, e^{imx}\right\rangle &= \int_{-\pi}^{\pi} e^{inx}\overline{e^{imx}}dx \\ &= \int_{-\pi}^{\pi} e^{inx}e^{-imx}dx \\ &= \frac{1}{i(n-m)}e^{i(n-m)x}\Big|_{-\pi}^{\pi} = 0.\end{aligned}$$
By dividing the functions in this set by
$$\left\|e^{inx}\right\| = \left[\int_{-\pi}^{\pi} e^{inx}\overline{e^{inx}}dx\right]^{1/2} = \sqrt{2\pi}, \quad n \in \mathbb{Z},$$
we obtain the corresponding orthonormal set
$$\left\{\frac{1}{\sqrt{2\pi}}e^{inx} : n \in \mathbb{Z}\right\}.$$

If ρ is a positive continuous function on (a,b), we define the inner product of two functions $f, g \in C(a,b)$ with respect to the weight function ρ by
$$\langle f, g\rangle_\rho = \int_a^b f(x)\bar{g}(x)\rho(x)dx, \tag{1.13}$$
and we leave it to the reader to verify that all the properties of the inner product, as given in Definition 1.5, are satisfied. f is then said to be orthogonal to g with respect to the weight function ρ if $\langle f, g\rangle_\rho = 0$. The induced norm
$$\|f\|_\rho = \left[\int_a^b |f(x)|^2 \rho(x)dx\right]^{1/2}$$
satisfies all the properties of the norm (1.11), including the CBS inequality and the triangle inequality. We use $\mathcal{L}^2_\rho(a,b)$ to denote the set of functions f :

$(a,b) \to \mathbb{C}$, where (a,b) may be finite or infinite, such that $\|f\|_\rho < \infty$. This is clearly an inner product space, and $\mathcal{L}^2(a,b)$ is then the special case in which $\rho \equiv 1$.

EXERCISES

1.19 Prove the triangle inequality $\|f+g\| \leq \|f\| + \|g\|$ for any $f, g \in \mathcal{L}^2(a,b)$.

1.20 Verify the CBS inequality for the functions $f(x) = 1$ and $g(x) = x$ on $[0,1]$.

1.21 Let the functions f and g be defined on $[0,1]$ by

$$f(x) = \begin{cases} 1, & x \in \mathbb{Q} \cap [0,1] \\ -1, & x \in \mathbb{Q}^c \cap [0,1] \end{cases}, \quad g(x) = 1 \quad \text{for all } x \in [0,1],$$

where $\mathbb{Q}$ is the set of rational numbers. Show that both f^2 and g^2 are Riemann integrable on $[0,1]$ but that fg is not.

1.22 Determine which of the following functions belongs to $\mathcal{L}^2(0,\infty)$ and calculate its norm.

$$(i)\ e^{-x}, \quad (ii)\ \sin x, \quad (iii)\ \frac{1}{1+x}, \quad (iv)\ \frac{1}{\sqrt[3]{x}}.$$

1.23 If f and g are positive, continuous functions in $\mathcal{L}^2(a,b)$, prove that $\langle f, g\rangle = \|f\| \, \|g\|$ if, and only if, f and g are linearly dependent.

1.24 Discuss the conditions under which the equality $\|f+g\| = \|f\| + \|g\|$ holds in $\mathcal{L}^2(a,b)$.

1.25 Determine the real values of α for which x^α lies in $\mathcal{L}^2(0,1)$.

1.26 Determine the real values of α for which x^α lies in $\mathcal{L}^2(1,\infty)$.

1.27 If $f \in \mathcal{L}^2(0,\infty)$ and $\lim_{x\to\infty} f(x)$ exists, prove that $\lim_{x\to\infty} f(x) = 0$.

1.28 Assuming that the interval (a,b) is finite, prove that if $f \in \mathcal{L}^2(a,b)$ then the integral $\int_a^b |f(x)|\,dx$ exists. Show that the converse is false by giving an example of a function f such that $|f|$ is integrable on (a,b), but $f \notin \mathcal{L}^2(a,b)$.

1.29 If the function $f : [0,\infty) \to \mathbb{R}$ is bounded and $|f|$ is integrable, prove that $f \in \mathcal{L}^2(0,\infty)$. Show that the converse is false by giving an example of a bounded function in $\mathcal{L}^2(0,\infty)$ which is not integrable on $[0,\infty)$.

1.30 In $\mathcal{L}^2(-\pi, \pi)$, express the function $\sin^3 x$ as a linear combination of the orthogonal functions $\{1, \cos x, \sin x, \cos 2x, \sin 2x, \ldots\}$.

1.31 Define a function $f \in \mathcal{L}^2(-1, 1)$ such that $\langle f, x^2 + 1 \rangle = 0$ and $\|f\| = 2$.

1.32 Given $\rho(x) = e^{-x}$, prove that any polynomial in x belongs to $\mathcal{L}^2_\rho(0, \infty)$.

1.33 Show that if ρ and σ are two weight functions such that $\rho \geq \sigma \geq 0$ on (a, b), then $\mathcal{L}^2_\rho(a, b) \subseteq \mathcal{L}^2_\sigma(a, b)$.

1.4 Sequences of Functions

Much of the subject of this book deals with sequences and series of functions, and this section presents the background that we need on their convergence properties. We assume that the reader is familiar with the basic theory of numerical sequences and series which is usually covered in advanced calculus.

Suppose that for each $n \in \mathbb{N}$ we have a (real or complex) function $f_n : I \to \mathbb{F}$ defined on a real interval I. We then say that we have a *sequence of functions* $(f_n : n \in \mathbb{N})$ defined on I. Suppose, furthermore, that, for every fixed $x \in I$, the *sequence of numbers* $(f_n(x) : n \in \mathbb{N})$ converges as $n \to \infty$ to some limit in $\mathbb{F}$. Now we define the function $f : I \to \mathbb{F}$, for each $x \in I$, by

$$f(x) = \lim_{n\to\infty} f_n(x). \tag{1.14}$$

That means, given any positive number ε, there is a positive integer N such that

$$n \geq N \Rightarrow |f_n(x) - f(x)| < \varepsilon. \tag{1.15}$$

Note that the number N depends on the point x as much as it depends on ε, hence $N = N(\varepsilon, x)$. The function f defined in Equation (1.14) is called the *pointwise limit* of the sequence (f_n).

Definition 1.13

A sequence of functions $f_n : I \to \mathbb{F}$ is said to *converge pointwise* to the function $f : I \to \mathbb{F}$, expressed symbolically by

$$\lim_{n\to\infty} f_n = f, \quad \lim f_n = f, \quad \text{or } f_n \to f,$$

if, for every $x \in I$, $\lim_{n\to\infty} f_n(x) = f(x)$.

Example 1.14

(i) Let $f_n(x) = \frac{1}{n}\sin nx$, $x \in \mathbb{R}$. In as much as

$$\lim_{n\to\infty} f_n(x) = \lim_{n\to\infty} \frac{1}{n}\sin nx = 0 \quad \text{for every } x \in \mathbb{R},$$

the pointwise limit of this sequence is the function $f(x) = 0,\ x \in \mathbb{R}$.

(ii) For all $x \in [0,1]$,

$$f_n(x) = x^n \to \begin{cases} 0, & 0 \le x < 1 \\ 1, & x = 1, \end{cases}$$

hence the limit function is

$$f(x) = \begin{cases} 0, & 0 \le x < 1 \\ 1, & x = 1, \end{cases} \tag{1.16}$$

as shown in Figure 1.1.

(iii) For all $x \in [0,\infty)$,

$$f_n(x) = \frac{nx}{1+nx} \to f(x) = \begin{cases} 0, & x = 0 \\ 1, & x > 0. \end{cases}$$

Example 1.15

For each $n \in \mathbb{N}$, define the sequence $f_n : [0,1] \to \mathbb{R}$ by

$$f_n(x) = \begin{cases} 0, & x = 0 \\ n, & 0 < x \le 1/n \\ 0, & 1/n < x \le 1. \end{cases}$$

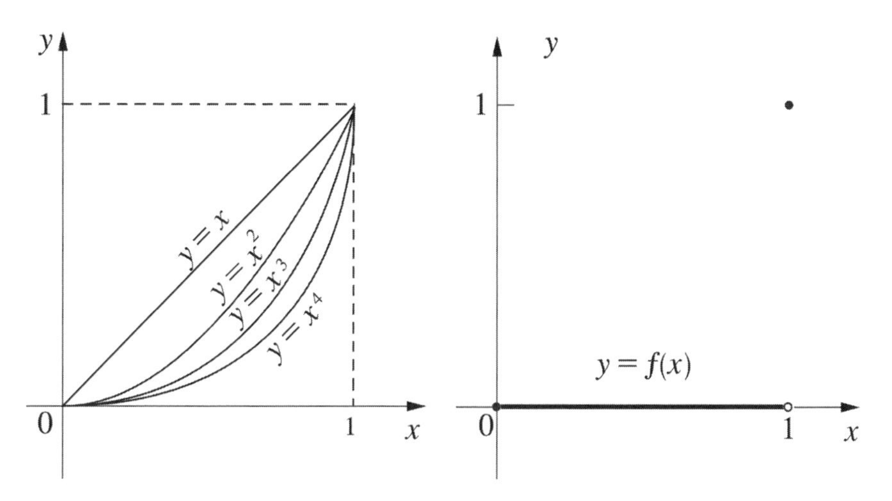

Figure 1.1 The sequence $f_n(x) = x^n$.

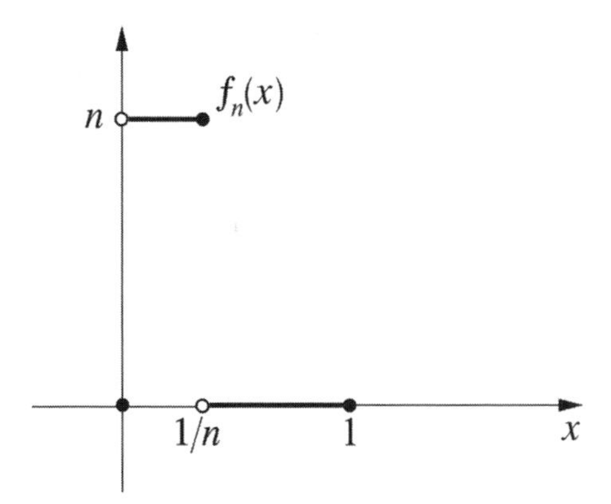

Figure 1.2

To determine the limit f, we first note that $f_n(0) = 0$ for all n. If $x > 0$, then there is an integer N such that $1/N < x$, in which case

$$n \geq N \Rightarrow \frac{1}{n} \leq \frac{1}{N} < x \Rightarrow f_n(x) = 0.$$

Therefore $f_n \to 0$ (see Figure 1.2).

If the number N in the implication (1.15) does not depend on x, that is, if for every $\varepsilon > 0$ there is an integer $N = N(\varepsilon)$ such that

$$n \geq N \Rightarrow |f_n(x) - f(x)| < \varepsilon \quad \text{for all } x \in I, \tag{1.17}$$

then the convergence $f_n \to f$ is called *uniform*, and we distinguish this from pointwise convergence by writing

$$f_n \xrightarrow{u} f.$$

Going back to Example 1.14, we note the following.

(i) Since

$$|f_n(x) - 0| = \left|\frac{1}{n} \sin nx\right| \leq \frac{1}{n} \quad \text{for all } x \in \mathbb{R},$$

we see that any choice of N greater than $1/\varepsilon$ will satisfy the implication (1.17), hence

$$\frac{1}{n} \sin nx \xrightarrow{u} 0 \ \text{ on } \mathbb{R}.$$

(ii) The convergence $x^n \to 0$ is not uniform on $[0, 1)$ because the implication

$$n \geq N \Rightarrow |x^n - 0| = x^n < \varepsilon$$

cannot be satisfied on the whole interval $[0, 1)$ if $0 < \varepsilon < 1$, but only on $[0, \sqrt[n]{\varepsilon})$, because $x^n > \varepsilon$ for all $x \in (\sqrt[n]{\varepsilon}, 1)$. Hence the convergence $f_n \to f$, where f is given in (1.16), is not uniform.

(iii) The convergence

$$\frac{nx}{1+nx} \to 1, \quad x \in (0, \infty)$$

is also not uniform in as much as the inequality

$$\left| \frac{nx}{1+nx} - 1 \right| = \frac{1}{1+nx} < \varepsilon$$

cannot be satisfied for values of x in $(0, (1-\varepsilon)/n\varepsilon]$ if $0 < \varepsilon < 1$.

Remark 1.16

1. The uniform convergence $f_n \xrightarrow{u} f$ clearly implies the pointwise convergence $f_n \to f$ (but not vice versa). Hence, when we wish to test for the uniform convergence of a sequence f_n, the candidate function f for the uniform limit of f_n should always be the pointwise limit.

2. In the inequalities (1.15) and (1.17) we can replace the relation $<$ by $\leq$ and the positive number ε by $c\varepsilon$, where c is a positive constant (which does not depend on n).

3. Because the statement $|f_n(x) - f(x)| \leq \varepsilon$ for all $x \in I$ is equivalent to

$$\sup_{x \in I} |f_n(x) - f(x)| \leq \varepsilon,$$

we see that $f_n \xrightarrow{u} f$ on I if, and only if, for every $\varepsilon > 0$ there is an integer N such that

$$n \geq N \Rightarrow \sup_{x \in I} |f_n(x) - f(x)| \leq \varepsilon,$$

which is equivalent to the statement

$$\sup_{x \in I} |f_n(x) - f(x)| \to 0 \quad \text{as } n \to \infty. \tag{1.18}$$

Using the criterion (1.18) for uniform convergence on the sequences of Example 1.14, we see that, in (i),

$$\sup_{x \in \mathbb{R}} \left| \frac{1}{n} \sin nx \right| \leq \frac{1}{n} \to 0,$$

thus confirming the uniform convergence of $\sin nx/n$ to 0. In (ii) and (iii), we have

$$\sup_{x \in [0,1]} |x^n - f(x)| = \sup_{x \in [0,1)} x^n = 1 \nrightarrow 0,$$

$$\sup_{x\in[0,\infty)}\left|\frac{nx}{1+nx}-f(x)\right| = \sup_{x\in(0,\infty)}\left(1-\frac{nx}{1+nx}\right) = 1 \nrightarrow 0,$$

hence neither sequence converges uniformly.

Although all three sequences discussed in Example 1.14 are continuous, only the first one, $(\sin nx/n)$, converges to a continuous limit. This would seem to indicate that uniform convergence preserves the property of continuity as the sequence passes to the limit. We should also be interested to know under what conditions we can interchange the operations of integration or differentiation with the process of passage to the limit. In other words, when can we write

$$\int_I \lim f_n(x)dx = \lim \int_I f_n(x)dx, \quad \text{or} \quad (\lim f_n)' = \lim\ f_n' \text{ on } I?$$

The answer is contained in the following theorem, which gives sufficient conditions for the validity of these equalities. This is a standard result in classical real analysis whose proof may be found in a number of references, such as [1] or [14].

Theorem 1.17

Let (f_n) be a sequence of functions defined on the interval I which converges pointwise to f on I.

(i) If f_n is continuous for every n, and $f_n \xrightarrow{u} f$, then f is continuous on I.

(ii) If f_n is integrable for every n, I is bounded, and $f_n \xrightarrow{u} f$, then f is integrable on I and

$$\int_I f(x)dx = \lim \int_I f_n(x)dx.$$

(iii) If f_n is differentiable on I for every n, I is bounded, and f_n' converges uniformly on I, then f_n converges uniformly to f, f is differentiable on I, and

$$f_n' \xrightarrow{u} f' \text{ on } I.$$

Remark 1.18

Part (iii) of Theorem 1.17 remains valid if pointwise convergence of f_n on I is replaced by the weaker condition that f_n converges at any single point in I, for such a condition is only needed to ensure the convergence of the constants of integration in going from f_n' to f_n.

Going back to Example 1.14, we observe that the uniform convergence of $\sin nx/n$ to 0 satisfies part (i) of Theorem 1.17. It also satisfies (ii) over any bounded interval in $\mathbb{R}$. But (iii) is not satisfied, in as much as the sequence

$$\frac{d}{dx}\left(\frac{1}{n}\sin nx\right) = \cos nx$$

is not convergent. The sequence (x^n) is continuous on $[0,1]$ for every n, but its limit is not. This is consistent with (i), because the convergence is not uniform. The same observation applies to the sequence $nx/(1+nx)$.

In Example 1.15 we have

$$\int_0^1 f_n(x)dx = \int_0^{1/n} ndx = 1 \quad \text{for all } n \in \mathbb{N}$$
$$\Rightarrow \lim \int_0^1 f_n(x)dx = 1,$$

whereas

$$\int_0^1 \lim f_n(x)dx = 0.$$

This implies that the convergence $f_n \to 0$ is not uniform, which is confirmed by the fact that

$$\sup_{0\leq x\leq 1} f_n(x) = n.$$

On the other hand,

$$\lim \int_0^1 x^n dx = 0 = \int_0^1 \lim x^n dx,$$

although the convergence $x^n \to 0$ is not uniform, which indicates that not all the conditions of Theorem 1.17 are necessary.

Given a sequence of (real or complex) functions (f_n) defined on a real interval I, we define its nth partial sum by

$$S_n(x) = f_1(x) + \cdots + f_n(x) = \sum_{k=1}^{n} f_k(x), \quad x \in I.$$

The sequence of functions (S_n), defined on I, is called an *infinite series* (of functions) and is denoted $\sum f_k$. The series is said to *converge pointwise* on I if the sequence (S_n) converges pointwise on I, in which case $\sum f_k$ is called *convergent*. Its limit is the sum of the series

$$\lim_{n\to\infty} S_n(x) = \sum_{k=1}^{\infty} f_k(x), \; x \in I.$$

Sometimes we shall find it convenient to identify a convergent series with its sum, just as we occasionally identify a function f with its value $f(x)$. A series which does not converge at a point is said to *diverge* at that point. The series $\sum f_k$ is *absolutely convergent* on I if the positive series $\sum |f_k|$ is pointwise convergent on I, and *uniformly convergent* on I if the sequence (S_n) is uniformly convergent on I. In investigating the convergence properties of series of functions we naturally rely on the corresponding convergence properties of sequences of functions, as discussed earlier, because a series is ultimately a sequence. But we shall often resort to the convergence properties of series of numbers, which we assume that the reader is familiar with, such as the various tests of convergence (comparison test, ratio test, root test, alternating series test), and the behaviour of such series as the geometric series and the p-series (see [1] or [3]).

Applying Theorem 1.17 to series, we arrive at the following result.

Corollary 1.19

Suppose the series $\sum f_n$ converges pointwise on the interval I.

(i) If f_n is continuous on I for every n and $\sum f_n$ converges uniformly on I, then its sum $\sum_{n=1}^{\infty} f_n$ is continuous.

(ii) If f_n is integrable on I for every n, I is bounded, and $\sum f_n$ converges uniformly, then $\sum_{n=1}^{\infty} f_n$ is integrable on I and

$$\int_I \sum_{n=1}^{\infty} f_n(x)dx = \sum_{n=1}^{\infty} \int_I f_n(x)dx.$$

(iii) If f_n is differentiable on I for every n, I is bounded, and $\sum f_n'$ converges uniformly on I, then $\sum f_n$ converges uniformly and its limit is differentiable on I and satisfies

$$\left(\sum_{n=1}^{\infty} f_n\right)' = \sum_{n=1}^{\infty} f_n'.$$

This corollary points out the relevance of uniform convergence to manipulating series, and it would be helpful if we had a simpler and more practical test for the uniform convergence of a series than applying the definition. This is provided by the following theorem, which gives a sufficient condition for the uniform convergence of a series of functions.

Theorem 1.20 (Weierstrass M-Test)

Let (f_n) be a sequence of functions on I, and suppose that there is a sequence of (nonnegative) numbers M_n such that

$$|f_n(x)| \leq M_n \quad \text{for all } x \in I, \quad n \in \mathbb{N}.$$

If $\sum M_n$ converges, then $\sum f_n$ converges uniformly and absolutely on I.

Proof

Let $\varepsilon > 0$. We have

$$\begin{aligned}\left|\sum_{k=1}^{\infty} f_k(x) - \sum_{k=1}^{n} f_k(x)\right| &\leq \sum_{k=n+1}^{\infty} |f_k(x)| \\ &\leq \sum_{k=n+1}^{\infty} M_k \quad \text{for all } x \in I, \quad n \in \mathbb{N}.\end{aligned}$$

Because the series $\sum M_k$ is convergent, there is an integer N such that

$$\begin{aligned} n \geq N &\Rightarrow \sum_{k=n+1}^{\infty} M_k < \varepsilon \\ &\Rightarrow \left|\sum_{k=1}^{\infty} f_k(x) - \sum_{k=1}^{n} f_k(x)\right| < \varepsilon \quad \text{for all } x \in I.\end{aligned}$$

By definition, this means $\sum f_k$ is uniformly convergent on I. Absolute convergence follows by comparison with M_n. □

Example 1.21

(i) The trigonometric series

$$\sum \frac{1}{n^2} \sin nx$$

is uniformly convergent on $\mathbb{R}$ because

$$\left|\frac{1}{n^2} \sin nx\right| \leq \frac{1}{n^2}$$

and the series $\sum 1/n^2$ is convergent. Because $\sin nx/n^2$ is continuous on $\mathbb{R}$ for every n, the function $\sum_{n=1}^{\infty} \sin nx/n^2$ is also continuous on $\mathbb{R}$. Furthermore, by Corollary 1.19, the integral of the series on any finite interval $[a, b]$ is

$$\begin{aligned}\int_a^b \left(\sum_{n=1}^{\infty} \frac{1}{n^2} \sin nx\right) dx &= \sum_{n=1}^{\infty} \frac{1}{n^2} \int_a^b \sin nx \; dx \\ &= \sum_{n=1}^{\infty} \frac{1}{n^3}(\cos na - \cos nb) \\ &\leq 2\sum_{n=1}^{\infty} \frac{1}{n^3},\end{aligned}$$

which is convergent. On the other hand, the series of derivatives

$$\sum_{n=1}^{\infty} \frac{d}{dx}\left(\frac{1}{n^2} \sin nx\right) = \sum_{n=1}^{\infty} \frac{1}{n} \cos nx$$

is not uniformly convergent. In fact, it is not even convergent at some values of x, such as the integral multiples of 2π. Hence we cannot write

$$\frac{d}{dx}\sum_{n=1}^{\infty}\frac{1}{n^2}\sin nx=\sum_{n=1}^{\infty}\frac{1}{n}\cos nx \quad \text{for all } x\in\mathbb{R}.$$

(ii) By the M-test, both the series

$$\sum\frac{1}{n^3}\sin nx$$

and

$$\sum_{n=1}^{\infty}\frac{d}{dx}\left(\frac{1}{n^3}\sin nx\right)=\sum_{n=1}^{\infty}\frac{1}{n^2}\cos nx$$

are uniformly convergent on $\mathbb{R}$. Hence the equality

$$\frac{d}{dx}\sum_{n=1}^{\infty}\frac{1}{n^3}\sin nx=\sum_{n=1}^{\infty}\frac{1}{n^2}\cos nx$$

is valid for all x in $\mathbb{R}$.

EXERCISES

1.34 Calculate the pointwise limit where it exists.

(a) $\dfrac{x^n}{1+x^n}$, $x\in\mathbb{R}$.

(b) $\sqrt[n]{x}$, $0\le x<\infty$.

(c) $\sin nx$, $x\in\mathbb{R}$.

1.35 Determine the type of convergence (pointwise or uniform) for each of the following sequences.

(a) $\dfrac{x^n}{1+x^n}$, $0\le x\le 2$.

(b) $\sqrt[n]{x}$, $1/2\le x\le 1$.

(c) $\sqrt[n]{x}$, $0\le x\le 1$.

1.36 Determine the type of convergence for the sequence

$$f_n(x)=\begin{cases} nx, & 0\le x<1/n\\ 1, & 1/n\le x\le 1,\end{cases}$$

and decide whether the equality

$$\lim\int_0^1 f_n(x)dx=\int_0^1\lim f_n(x)dx$$

is valid.

1.37 Evaluate the limit of the sequence

$$f_n(x) = \begin{cases} nx, & 0 \leq x \leq 1/n \\ n(1-x)/(n-1), & 1/n < x \leq 1, \end{cases}$$

and determine the type of convergence.

1.38 Determine the limit and the type of convergence for the sequence $f_n(x) = nx(1-x^2)^n$ on $[0,1]$.

1.39 Prove that the convergence

$$\frac{x}{n+x} \to 0$$

is uniform on $[0,a]$ for any $a > 0$, but not on $[0,\infty)$.

1.40 Given

$$f_n(x) = \begin{cases} 1/n, & |x| < n \\ 0, & |x| > n, \end{cases}$$

prove that $f_n \xrightarrow{u} 0$. Evaluate $\lim \int_{-\infty}^{\infty} f_n(x)dx$ and explain why it is not 0.

1.41 If the sequence (f_n) converges uniformly to f on $[a,b]$, prove that $|f_n - f|$, and hence $|f_n - f|^2$, converges uniformly to 0 on $[a,b]$.

1.42 Determine the domain of convergence of the series $\sum f_n$, where

(a) $f_n(x) = \dfrac{1}{n^2 + x^2}$.

(b) $f_n(x) = \dfrac{x^n}{1 + x^n}$.

1.43 If the series $\sum a_n$ is absolutely convergent, prove that $\sum a_n \sin nx$ is uniformly convergent on $\mathbb{R}$.

1.44 Prove that

$$\lim_{n\to\infty} \int_{n\pi}^{(n+1)\pi} \frac{|\sin x|}{x} dx = 0.$$

Use this to conclude that the improper integral

$$\int_0^{\infty} \frac{\sin x}{x} dx$$

exists. Show that the integral $\int_0^{\infty} (|\sin x|/x)dx = \infty$. Hint: Use the alternating series test and the divergence of the harmonic series $\sum 1/n$.

1.45 The series

$$\sum_{n=0}^{\infty} a_n x^n = a_0 + a_1 x + a_2 x^2 + \cdots$$

is called a *power series* about the point 0. It is known (see [1]) that this series converges in $(-R, R)$ and diverges outside $[-R, R]$, where

$$R = \left[\lim_{n\to\infty} \sqrt[n]{|a_n|}\right]^{-1} = \lim_{n\to\infty} \left|\frac{a_n}{a_{n+1}}\right| \geq 0.$$

If $R > 0$, use the Weierstrass M-test to prove that the power series converges uniformly on $[-R+\varepsilon, R-\varepsilon]$, where ε is any positive number less than R.

1.46 Use the result of Exercise 1.45 to show that the function

$$f(x) = \sum_{n=0}^{\infty} a_n x^n$$

is continuous on $(-R, R)$; then show that f is also differentiable on $(-R, R)$ with

$$f'(x) = \sum_{n=1}^{\infty} n a_n x^{n-1}.$$

1.47 From Exercise 1.46 conclude that the power series $f(x) = \sum_{n=0}^{\infty} a_n x^n$ is differentiable any number of times on $(-R, R)$, and that $a_n = f^{(n)}(0)/n!$ for all $n \in \mathbb{N}$.

1.48 Use the result of Exercise 1.47 to obtain the following power series (Taylor series) representations of the exponential and trigonometric functions on $\mathbb{R}$.

$$e^x = \sum_{n=0}^{\infty} \frac{x^n}{n!},$$

$$\cos x = \sum_{n=0}^{\infty} (-1)^n \frac{x^{2n}}{(2n)!},$$

$$\sin x = \sum_{n=0}^{\infty} (-1)^n \frac{x^{2n+1}}{(2n+1)!}.$$

1.49 Use the result of Exercise 1.48 to prove Euler's formula $e^{ix} = \cos x + i \sin x$ for all $x \in \mathbb{R}$, where $i = \sqrt{-1}$.

1.5 Convergence in $\mathcal{L}^2$

Having discussed pointwise and uniform convergence for a sequence of functions, we now consider a third type: convergence in $\mathcal{L}^2$.

Definition 1.22

A sequence of functions (f_n) in $\mathcal{L}^2(a,b)$ is said to *converge in* $\mathcal{L}^2$ if there is a function $f \in \mathcal{L}^2(a,b)$ such that

$$\lim_{n\to\infty} \|f_n - f\| = 0, \tag{1.19}$$

that is, if for every $\varepsilon > 0$ there is an integer N such that

$$n \geq N \Rightarrow \|f_n - f\| < \varepsilon.$$

Equation (1.19) is equivalent to writing

$$f_n \xrightarrow{\mathcal{L}^2} f,$$

and f is called the *limit in* $\mathcal{L}^2$ of the sequence (f_n).

Example 1.23

(i) In Example 1.14(ii) we saw that, pointwise,

$$x^n \to \begin{cases} 0, & 0 \leq x < 1 \\ 1, & x = 1. \end{cases}.$$

Because $\mathcal{L}^2([0,1]) = \mathcal{L}^2([0,1))$, we have

$$\|x^n - 0\| = \left[\int_0^1 x^{2n} dx\right]^{1/2} = \left[\frac{1}{2n+1}\right]^{1/2} \to 0.$$

Therefore $x^n \xrightarrow{\mathcal{L}^2} 0$.

(ii) The sequence of functions (f_n) defined in Example 1.15 by

$$f_n(x) = \begin{cases} 0, & x = 0 \\ n, & 0 < x \leq 1/n \\ 0, & 1/n < x \leq 1 \end{cases}$$

also converges pointwise to 0 on $[0,1]$. But in this case,

$$\begin{aligned}\|f_n - 0\|^2 &= \int_0^1 f_n^2(x)dx \\ &= \int_0^{1/n} n^2 dx \\ &= n \quad \text{for all } n \in \mathbb{N}.\end{aligned}$$

Thus $\|f_n - 0\| = \sqrt{n} \nrightarrow 0$, which means the sequence f_n does not converge to 0 in $\mathcal{L}^2$.

This last example shows that pointwise convergence does not imply convergence in $\mathcal{L}^2$. Conversely, convergence in $\mathcal{L}^2$ cannot imply pointwise convergence, because the limit in this case is a class of functions (which are equal in $\mathcal{L}^2$ but not pointwise). It is legitimate to ask, however, whether a sequence that converges pointwise to some limit f can converge to a different limit in $\mathcal{L}^2$. For example, can the sequence (f_n) in Example 1.23(ii) converge in $\mathcal{L}^2$ to some function other than 0? The answer is no. In other words, if a sequence converges both pointwise and in $\mathcal{L}^2$, then its limit is the same in both cases. More precisely, we should say that the two limits are not distinguishable in $\mathcal{L}^2$ as they belong to the same equivalence class.

On the other hand, uniform convergence $f_n \xrightarrow{u} f$ over I implies pointwise convergence, as we have already observed, and we now show that it also implies $f_n \xrightarrow{\mathcal{L}^2} f$ provided the sequence (f_n) and f lie in $\mathcal{L}^2(I)$ and I is bounded: Because $f_n - f \xrightarrow{u} 0$, it is a simple matter to show that $|f_n - f|^2 \xrightarrow{u} 0$ (Exercise 1.41). By Theorem 1.17(ii), we therefore have

$$\begin{aligned}\lim_{n\to\infty} \|f_n - f\|^2 &= \lim_{n\to\infty} \int_I |f_n(x) - f(x)|^2\, dx \\ &= \int_I \lim_{n\to\infty} |f_n(x) - f(x)|^2\, dx = 0.\end{aligned}$$

The condition that f belong to $\mathcal{L}^2(I)$ is actually not needed, as we shall discover in Theorem 1.26.

Example 1.24

We saw in Example 1.21 that

$$S_n(x) = \sum_{k=1}^{n} \frac{1}{k^2} \sin kx \xrightarrow{u} S(x) = \sum_{k=1}^{\infty} \frac{1}{k^2} \sin kx, \quad x \in \mathbb{R},$$

hence the function $S(x)$ is continuous on $[-\pi, \pi]$. Moreover, both S_n and S lie in $\mathcal{L}^2(-\pi, \pi)$ because each is uniformly bounded above by the convergent series

$\sum 1/k^2$. Therefore S_n converges to S in $\mathcal{L}^2(-\pi,\pi)$. Equivalently, we say that the series $\sum \sin kx/k^2$ converges to $\sum_{k=1}^{\infty} \sin kx/k^2$ in $\mathcal{L}^2(-\pi,\pi)$ and write

$$\lim \sum_{k=1}^{n} \frac{1}{k^2} \sin kx = \sum_{k=1}^{\infty} \frac{1}{k^2} \sin kx \text{ in } \mathcal{L}^2(-\pi,\pi).$$

The series $\sum \sin kx/k$, on the other hand, cannot be tested for convergence in $\mathcal{L}^2$ with the tools available, and we have to develop the theory a little further. First we define a Cauchy sequence in $\mathcal{L}^2$ along the lines of the corresponding notion in $\mathbb{R}$. This allows us to test a sequence for convergence without having to guess its limit beforehand.

Definition 1.25

A sequence in $\mathcal{L}^2$ is called a *Cauchy sequence* if, for every $\varepsilon > 0$, there is an integer N such that

$$m,n \geq N \Rightarrow \|f_n - f_m\| < \varepsilon.$$

Clearly, every convergent sequence (f_n) in $\mathcal{L}^2$ is a Cauchy sequence; for if $f_n \stackrel{\mathcal{L}^2}{\to} f$, then, by the triangle inequality,

$$\|f_n - f_m\| \leq \|f_n - f\| + \|f_m - f\|,$$

and we can make the right-hand side of this inequality arbitrarily small by taking m and n large enough. The converse of this statement (i.e., that every Cauchy sequence in $\mathcal{L}^2$ converges to some function in $\mathcal{L}^2$) is also true and expresses the *completeness property* of $\mathcal{L}^2$.

Theorem 1.26 (Completeness of $\mathcal{L}^2$)

For every Cauchy sequence (f_n) in $\mathcal{L}^2$ there is a function $f \in \mathcal{L}^2$ such that $f_n \stackrel{\mathcal{L}^2}{\to} f$.

There is another theorem which states that, for every function $f \in \mathcal{L}^2(a,b)$, there is a sequence of continuous functions (f_n) on $[a,b]$ such that $f_n \stackrel{\mathcal{L}^2}{\to} f$. In other words, the set of functions $C([a,b])$ is dense in $\mathcal{L}^2(a,b)$ in much the same way that the rationals $\mathbb{Q}$ are dense in $\mathbb{R}$, keeping in mind of course the different topologies of $\mathbb{R}$ and $\mathcal{L}^2$, the first being defined by the absolute value $|\cdot|$ and the second by the norm $\|\cdot\|$. For example, the $\mathcal{L}^2(-1,1)$ function

$$f(x) = \begin{cases} 0, & -1 \leq x < 0 \\ 1, & 0 \leq x \leq 1, \end{cases}$$

which is discontinuous at $x = 0$, can be approached in the $\mathcal{L}^2$ norm by the sequence of continuous functions

$$f_n = \begin{cases} 0, & -1 \leq x \leq -1/n \\ nx+1, & -1/n < x < 0 \\ 1, & 0 \leq x \leq 1. \end{cases}$$

This is clear from

$$\begin{aligned} \lim_{n\to\infty} \|f_n - f\| &= \lim_{n\to\infty} \left[\int_{-1}^{1} |f_n(x) - f(x)|^2 \, dx \right]^{1/2} \\ &= \lim_{n\to\infty} \left[\int_{-1/n}^{0} (nx+1)^2 dx \right]^{1/2} \\ &= \lim_{n\to\infty} 1/\sqrt{3n} = 0. \end{aligned}$$

Needless to say, there are many other sequences in $C([-1,1])$ which converge to f in $\mathcal{L}^2(-1,1)$, just as there are many sequences in $\mathbb{Q}$ which converge to the irrational number $\sqrt{2}$.

As we shall have occasion to refer to this result in the following chapter, we give here its precise statement.

Theorem 1.27 (Density of C in $\mathcal{L}^2$)

For any $f \in \mathcal{L}^2(a,b)$ and any $\varepsilon > 0$, there is a continuous function g on $[a,b]$ such that $\|f - g\| < \varepsilon$.

The proofs of Theorems 1.26 and 1.27 may be found in [14]. The space $\mathcal{L}^2$ is one of the most important examples of a *Hilbert space,* which is an inner product space that is complete under the norm defined by the inner product. It is named after David Hilbert (1862–1943), the German mathematician whose work and inspiration did much to develop the ideas of Hilbert space (see [7], vol. I). Many of the ideas that we work with are articulated within the context of $\mathcal{L}^2$.

Example 1.28

Using Theorem 1.26, we can now look into the question of convergence of the sequence $S_n(x) = \sum_{k=1}^{n} \sin kx/k$ in $\mathcal{L}^2(-\pi, \pi)$. Noting that

$$\|S_n(x) - S_m(x)\|^2 = \left\| \sum_{k=m+1}^{n} \frac{1}{k} \sin kx \right\|^2, \quad m < n,$$

we can use the orthogonality of $\{\sin kx : k \in \mathbb{N}\}$ in $\mathcal{L}^2(-\pi,\pi)$ (Example 1.11) to obtain

$$\left\| \sum_{k=m+1}^{n} \frac{1}{k} \sin kx \right\|^2 = \sum_{k=m+1}^{n} \frac{1}{k^2} \|\sin kx\|^2 = \pi \sum_{k=m+1}^{n} \frac{1}{k^2}.$$

Suppose $\varepsilon > 0$. Since $\sum 1/k^2$ is convergent, we can choose N so that

$$\begin{aligned} n > m \geq N &\Rightarrow \sum_{k=m+1}^{n} \frac{1}{k^2} < \frac{\varepsilon^2}{\pi} \\ &\Rightarrow \|S_n(x) - S_m(x)\| < \varepsilon. \end{aligned}$$

Thus $\sum_{k=1}^{n} \sin kx/k$ is a Cauchy sequence and hence converges in $\mathcal{L}^2(-\pi,\pi)$, although we cannot as yet tell to what limit.

Similarly, the series $\sum \cos kx/k$ converges in $\mathcal{L}^2(-\pi,\pi)$, although this series diverges pointwise at certain values of x, such as all integral multiples of 2π.

This section was devoted to convergence in $\mathcal{L}^2$ because of its importance to the theory of Fourier series, but we could just as easily have been discussing convergence in the weighted space $\mathcal{L}^2_\rho$. Definitions 1.22 and 1.25 and Theorems 1.26 and 1.27 would remain unchanged, with the norm $\|\cdot\|$ replaced by $\|\cdot\|_\rho$ and convergence in $\mathcal{L}^2$ by convergence in $\mathcal{L}^2_\rho$.

EXERCISES

1.50 Determine the limit in $\mathcal{L}^2$ of each of the following sequences where it exists.

(a) $f_n(x) = \sqrt[n]{x}$, $0 \leq x \leq 1$.

(b) $f_n(x) = \begin{cases} nx, & 0 \leq x < 1/n \\ 1, & 1/n \leq x \leq 1. \end{cases}$

(c) $f_n(x) = nx(1-x)^n$, $0 \leq x \leq 1$.

1.51 Test the following series for convergence in $\mathcal{L}^2$.

(a) $\sum \frac{1}{k^{2/3}} \sin kx$, $-\pi \leq x \leq \pi$.

(b) $\sum \frac{1}{k} e^{ikx}$, $-\pi \leq x \leq \pi$.

(c) $\sum \frac{1}{\sqrt{k+1}} \cos kx$, $-\pi \leq x \leq \pi$.

1.52 If (f_n) is a sequence in $\mathcal{L}^2(a,b)$ which converges to f in $\mathcal{L}^2$, show that $\langle f_n, g\rangle \xrightarrow{\mathcal{L}^2} \langle f, g\rangle$ for any $g \in \mathcal{L}^2(a,b)$.

1.53 Prove that $|\|f\| - \|g\|| \leq \|f - g\|$, and hence conclude that if $f_n \xrightarrow{\mathcal{L}^2} f$ then $\|f_n\| \to \|f\|$.

1.54 If the numerical series $\sum |a_n|$ is convergent, prove that $\sum |a_n|^2$ is also convergent, and that the series $\sum a_n \sin nx$ and $\sum a_n \cos nx$ are both continuous on $[-\pi, \pi]$.

1.55 Prove that if the weight functions ρ and σ are related by $\rho \geq \sigma$ on (a,b), then a sequence which converges in $\mathcal{L}^2_\rho(a,b)$ also converges in $\mathcal{L}^2_\sigma(a,b)$.

1.6 Orthogonal Functions

Let

$$\{\varphi_1, \varphi_2, \varphi_3, \ldots\}$$

be an orthogonal set of (nonzero) functions in the complex space $\mathcal{L}^2$, which may be finite or infinite, and suppose that the function $f \in \mathcal{L}^2$ is a finite linear combination of elements in the set $\{\varphi_i\}$,

$$f = \sum_{i=1}^{n} \alpha_i \varphi_i, \quad \alpha_i \in \mathbb{C}. \tag{1.20}$$

Taking the inner product of f with φ_k,

$$\langle f, \varphi_k\rangle = \alpha_k \|\varphi_k\|^2 \quad \text{for all } k = 1, \ldots, n,$$

we conclude that

$$\alpha_k = \frac{\langle f, \varphi_k\rangle}{\|\varphi_k\|^2},$$

and the representation (1.20) takes the form

$$f = \sum_{k=1}^{n} \frac{\langle f, \varphi_k\rangle}{\|\varphi_k\|^2} \varphi_k.$$

In other words, the coefficients α_k in the linear combination (1.20) are determined by the projections of f on φ_k. In terms of the corresponding orthonormal set $\{\psi_k = \varphi_k / \|\varphi_k\|\}$,

$$f = \sum_{k=1}^{n} \langle f, \psi_k\rangle \psi_k,$$

and the coefficients coincide with the projections of f on ψ_k.

Suppose, on the other hand, that f is an arbitrary function in $\mathcal{L}^2$ and that we want to obtain the best approximation of f in $\mathcal{L}^2$, that is, in the norm $\|\cdot\|$, by a finite linear combination of the elements of $\{\varphi_k\}$. We should then look for the coefficients α_k which minimize the nonnegative number

$$\left\| f - \sum_{k=1}^{n} \alpha_k \varphi_k \right\|.$$

We have

$$\begin{aligned}
\left\| f - \sum_{k=1}^{n} \alpha_k \varphi_k \right\|^2 &= \left\langle f - \sum_{k=1}^{n} \alpha_k \varphi_k, f - \sum_{k=1}^{n} \alpha_k \varphi_k \right\rangle \\
&= \|f\|^2 - 2 \sum_{k=1}^{n} \operatorname{Re} \bar{\alpha}_k \langle f, \varphi_k \rangle + \sum_{k=1}^{n} |\alpha_k|^2 \|\varphi_k\|^2 \\
&= \|f\|^2 - \sum_{k=1}^{n} \frac{|\langle f, \varphi_k \rangle|^2}{\|\varphi_k\|^2} \\
&\quad + \sum_{k=1}^{n} \|\varphi_k\|^2 \left[|\alpha_k^2| - 2 \operatorname{Re} \bar{\alpha}_k \frac{\langle f, \varphi_k \rangle}{\|\varphi_k\|^2} + \frac{|\langle f, \varphi_k \rangle|^2}{\|\varphi_k\|^4} \right] \\
&= \|f\|^2 - \sum_{k=1}^{n} \frac{|\langle f, \varphi_k \rangle|^2}{\|\varphi_k\|^2} + \sum_{k=1}^{n} \|\varphi_k\|^2 \left| \alpha_k - \frac{\langle f, \varphi_k \rangle}{\|\varphi_k\|^2} \right|^2 .
\end{aligned}$$

Since the coefficients α_k appear only in the last term

$$\sum_{k=1}^{n} \|\varphi_k\|^2 \left| \alpha_k - \frac{\langle f, \varphi_k \rangle}{\|\varphi_k\|^2} \right|^2 \geq 0,$$

we obviously achieve the minimum of $\|f - \sum_{k=1}^{n} \alpha_k \varphi_k\|^2$, and hence of $\|f - \sum_{k=1}^{n} \alpha_k \varphi_k\|$, by choosing

$$\alpha_k = \frac{\langle f, \varphi_k \rangle}{\|\varphi_k\|^2}.$$

This minimum is given by

$$\left\| f - \sum_{k=1}^{n} \frac{\langle f, \varphi_k \rangle}{\|\varphi_k\|^2} \varphi_k \right\|^2 = \|f\|^2 - \sum_{k=1}^{n} \frac{|\langle f, \varphi_k \rangle|^2}{\|\varphi_k\|^2} \geq 0. \tag{1.21}$$

This yields the relation,

$$\sum_{k=1}^{n} \frac{|\langle f, \varphi_k \rangle|^2}{\|\varphi_k\|^2} \leq \|f\|^2 .$$

Since this relation is true for any n, it is also true in the limit as $n \to \infty$. The resulting inequality

$$\sum_{k=1}^{\infty} \frac{|\langle f, \varphi_k \rangle|^2}{\|\varphi_k\|^2} \leq \|f\|^2, \tag{1.22}$$

known as *Bessel's inequality*, holds for any orthogonal set $\{\varphi_k : k \in \mathbb{N}\}$ and any $f \in \mathcal{L}^2$.

In view of (1.21), Bessel's inequality becomes an equality if, and only if,

$$\left\| f - \sum_{k=1}^{\infty} \frac{\langle f, \varphi_k \rangle}{\|\varphi_k\|^2} \varphi_k \right\| = 0,$$

or, equivalently,

$$f = \sum_{k=1}^{\infty} \frac{\langle f, \varphi_k \rangle}{\|\varphi_k\|^2} \varphi_k \text{ in } \mathcal{L}^2,$$

which means that f is represented in $\mathcal{L}^2$ by the sum $\sum_{k=1}^{\infty} \alpha_k \varphi_k$, where $\alpha_k = \langle f, \varphi_k \rangle / \|\varphi_k\|^2$.

Definition 1.29

An orthogonal set $\{\varphi_n : n \in \mathbb{N}\}$ in $\mathcal{L}^2$ is said to be *complete* if, for any $f \in \mathcal{L}^2$,

$$\sum_{k=1}^{n} \frac{\langle f, \varphi_k \rangle}{\|\varphi_k\|^2} \varphi_k \xrightarrow{\mathcal{L}^2} f.$$

Thus a complete orthogonal set in $\mathcal{L}^2$ becomes a basis for the space, and because $\mathcal{L}^2$ is infinite-dimensional the basis has to be an infinite set. When Bessel's inequality becomes an equality, the resulting relation

$$\|f\|^2 = \sum_{n=1}^{\infty} \frac{|\langle f, \varphi_n \rangle|^2}{\|\varphi_n\|^2} \tag{1.23}$$

is called *Parseval's relation* or the *completeness relation.* The second term is justified by the following theorem, which is really a restatement of Definition 1.29.

Theorem 1.30

An orthogonal set $\{\varphi_n : n \in \mathbb{N}\}$ is complete if, and only if, it satisfies Parseval's relation (1.23) for any $f \in \mathcal{L}^2$.

Remark 1.31

1. Given any orthogonal set $\{\varphi_n : n \in \mathbb{N}\}$ in $\mathcal{L}^2$, we have shown that we obtain the best $\mathcal{L}^2$-approximation

$$\sum_{k=1}^{n} \alpha_k \varphi_k$$

of the function $f \in \mathcal{L}^2$ by choosing $\alpha_k = \langle f, \varphi_k \rangle / \|\varphi_k\|^2$, and this choice is independent of n. If $\{\varphi_n\}$ is complete then the equality $f = \sum_{n=1}^{\infty} \alpha_n \varphi_n$ holds in $\mathcal{L}^2$.

2. When the orthogonal set $\{\varphi_n\}$ is normalized to $\{\psi_k = \varphi_k / \|\varphi_k\|\}$, Bessel's inequality takes the form

$$\sum_{k=1}^{\infty} |\langle f, \psi_k \rangle|^2 \leq \|f\|^2 ,$$

and Parseval's relation becomes

$$\|f\|^2 = \sum_{n=1}^{\infty} |\langle f, \psi_n \rangle|^2 .$$

3. For any $f \in \mathcal{L}^2$, because $\|f\| < \infty$, we conclude from Bessel's inequality that $\langle f, \psi_n \rangle \to 0$ whether the orthonormal set $\{\psi_n\}$ is complete or not.

Parseval's relation may be regarded as a generalization of the theorem of Pythagoras from $\mathbb{R}^n$ to $\mathcal{L}^2$, where $\|f\|^2$ replaces the square of the length of the vector, and $\sum_{n=1}^{\infty} |\langle f, \psi_n \rangle|^2$ represents the sum of the squares of its projections on the orthonormal basis. That is one reason why $\mathcal{L}^2$ is considered the natural generalization of the finite-dimensional Euclidean space to infinite dimensions. It preserves some of the basic geometric structure of $\mathbb{R}^n$, and the completeness property (Theorem 1.26) guarantees its closure under limiting operations on Cauchy sequences.

EXERCISES

1.56 If l is any positive number, show that $\{\sin(n\pi x/l) : n \in \mathbb{N}\}$ and $\{\cos(n\pi x/l) : n \in \mathbb{N}_0\}$ are orthogonal sets in $\mathcal{L}^2(0, l)$. Determine the corresponding orthonormal sets.

1.57 Determine the coefficients c_i in the linear combination

$$c_1 + c_2 \sin \pi x + c_3 \sin 2\pi x$$

which give the best approximation in $\mathcal{L}^2(0, 2)$ of the function $f(x) = x,\ 0 < x < 2$.

1.58 Determine the coefficients a_i and b_i in the linear combination

$$a_0 + a_1 \cos x + b_1 \sin x + a_2 \cos 2x + b_2 \sin 2x$$

which give the best approximation in $\mathcal{L}^2(-\pi, \pi)$ of $f(x) = |x|$, $-\pi \leq x \leq \pi$.

1.59 Let p_1, p_2, and p_3 be the three orthogonal polynomials formed from the set $\{1, x, x^2\}$ by the Gram–Schmidt method, where $-1 \leq x \leq 1$. Determine the constant coefficients in the second-degree polynomial $a_1 p_1(x) + a_2 p_2(x) + a_3 p_3(x)$ which give the best approximation in $\mathcal{L}^2(-1, 1)$ of e^x. Can you think of another polynomial p of degree 2 which approximates e^x on $(-1, 1)$ in a different sense?

1.60 Assuming that

$$1 - x = \frac{8}{\pi^2} \sum_{n=1}^{\infty} \frac{1}{(2n-1)^2} \cos \frac{(2n-1)\pi}{2} x, \quad 0 \leq x \leq 2,$$

use Parseval's identity to prove that

$$\pi^4 = 96 \sum_{n=1}^{\infty} \frac{1}{(2n-1)^4}.$$

1.61 Define a real sequence (a_k) such that $\sum a_k^2$ converges and $\sum a_k$ diverges. What type of convergence can the series $\sum a_n \cos nx$, $-\pi \leq x \leq \pi$ have?

1.62 Suppose $\{f_n : n \in \mathbb{N}\}$ is an orthogonal set in $\mathcal{L}^2(0, l)$, and let

$$\varphi_n(x) = \frac{1}{2}[f_n(x) + f_n(-x)],$$
$$\psi_n(x) = \frac{1}{2}[f_n(x) - f_n(-x)], \quad -l \leq x \leq l,$$

be the even and odd extensions, respectively, of f_n from $[0, l]$ to $[-l, l]$. Show that the set $\{\varphi_n\} \cup \{\psi_n\}$ is orthogonal in $\mathcal{L}^2(-l, l)$. If $\{f_n\}$ is orthonormal in $\mathcal{L}^2(0, l)$, what is the corresponding orthonormal set in $\mathcal{L}^2(-l, l)$?

2
The Sturm–Liouville Theory

Complete orthogonal sets of functions in $\mathcal{L}^2$ arise naturally as solutions of certain second-order linear differential equations under appropriate boundary conditions, commonly referred to as Sturm–Liouville boundary-value problems, after the Swiss mathematician Jacques Sturm (1803–1855) and the French mathematician Joseph Liouville (1809–1882), who studied these problems and the properties of their solutions. The differential equations considered here arise directly as mathematical models of motion according to Newton's law, but more often as a result of using the method of separation of variables to solve the classical partial differential equations of physics, such as Laplace's equation, the heat equation, and the wave equation.

2.1 Linear Second-Order Equations

Consider the ordinary differential equation of second order on the real interval I given by

$$a_0(x)y'' + a_1(x)y' + a_2(x)y = f(x), \tag{2.1}$$

where a_0, a_1, a_2, and f are given complex functions on I. When $f = 0$ on I, the equation is called *homogeneous*, otherwise it is *nonhomogeneous*. Any (complex) function $\varphi \in C^2(I)$ is a *solution* of Equation (2.1) if the substitution of y by φ results in the identity

$$a_0(x)\varphi''(x) + a_1(x)\varphi'(x) + a_2(x)\varphi(x) = f(x) \qquad \text{for all } x \in I.$$

If we denote the second-order *differential operator*

$$a_0(x)\frac{d^2}{dx^2} + a_1(x)\frac{d}{dx} + a_2(x)$$

by L, then Equation (2.1) can be written in the form $Ly = f$. The operator L is linear, in the sense that

$$L(c_1\varphi + c_2\psi) = c_1 L\varphi + c_2 L\psi$$

for any functions $\varphi, \psi \in C^2(I)$ and any constants $c_1, c_2 \in \mathbb{C}$, hence (2.1) is called a *linear differential equation.* Unless otherwise specified, all differential equations and operators that we deal with are linear. A fundamental property of linear homogeneous equations is that any linear combination of solutions of the equation is also a solution; for if φ and ψ satisfy

$$L\varphi = 0, \qquad L\psi = 0,$$

then we clearly have

$$L(c_1\varphi + c_2\psi) = c_1 L\varphi + c_2 L\psi = 0$$

for any pair of constants c_1 and c_2. This is known as the *superposition principle.*

If the function a_0 does not vanish at any point on I, Equation (2.1) may be divided by a_0 to give

$$y'' + q(x)y' + r(x)y = g(x), \tag{2.2}$$

where $q = a_1/a_0$, $r = a_2/a_0$, and $g = f/a_0$. Equations (2.1) and (2.2) are clearly equivalent, in the sense that they have the same set of solutions. Equation (2.1) is then said to be *regular* on I; otherwise, if there is a point $c \in I$ where $a_0(c) = 0$, the equation is *singular*, and c is then referred to as a *singular point* of the equation.

According to the existence and uniqueness theorem for linear equations (see [6]), if the functions q, r, and g are all continuous on I and x_0 is a point in I, then, for any two numbers ξ and η, there is a unique solution φ of (2.2) on I such that

$$\varphi(x_0) = \xi, \qquad \varphi'(x_0) = \eta. \tag{2.3}$$

Equations (2.3) are called *initial conditions*, and the system of equations (2.2) and (2.3) is called an *initial-value problem.*

Here we list some well-known properties of the solutions of Equation (2.2), which may be found in many standard introductions to ordinary differential equations.

1. The homogeneous equation

$$y'' + q(x)y' + r(x)y = 0, \qquad x \in I, \tag{2.4}$$

has two linearly independent solutions $y_1(x)$ and $y_2(x)$ on I. A linear combination of these two solutions

$$c_1 y_1 + c_2 y_2, \tag{2.5}$$

where c_1 and c_2 are arbitrary constants, is the *general solution* of (2.4); that is, any solution of the equation is given by (2.5) for some values of c_1 and c_2. When $c_1 = c_2 = 0$ we obtain the so-called *trivial solution* 0, which is always a solution of the homogeneous equation. By the uniqueness theorem, it is the only solution if $\xi = \eta = 0$ in (2.3).

2. If $y_p(x)$ is any particular solution of the nonhomogeneous Equation (2.2), then

$$y_p + c_1 y_1 + c_2 y_2$$

is the general solution of (2.2). By applying the initial conditions (2.3), the constants c_1 and c_2 are determined and we obtain the unique solution of the system of Equations (2.2) and (2.3).

3. When the coefficients q and r are constants, the general solution of Equation (2.4) has the form

$$c_1 e^{m_1 x} + c_2 e^{m_2 x},$$

where m_1 and m_2 are the roots of the second degree equation $m^2 + qm + r = 0$ when the roots are distinct. If $m_1 = m_2 = m$, then the solution takes the form $c_1 e^{mx} + c_2 x e^{mx}$, in which the functions e^{mx} and xe^{mx} are clearly linearly independent.

4. When $a_0(x) = x^2$, $a_1(x) = ax$, and $a_2(x) = b$, where a and b are constants, the homogeneous version of Equation (2.1) becomes

$$x^2 y'' + axy' + by = 0,$$

which is called the *Cauchy–Euler equation.* Its general solution is given by

$$c_1 x^{m_1} + c_2 x^{m_2},$$

where m_1 and m_2 are the distinct roots of $m^2 + (a-1)m + b = 0$. When $m_1 = m_2 = m$, the solution is $c_1 x^m + c_2 x^m \log x$.

5. If the coefficients q and r are *analytic functions* at some point x_0 in the interior of I, which means each may be represented in an open interval

centered at x_0 by a power series in $(x - x_0)$, then the general solution of (2.4) is also analytic at x_0, and is represented by a power series of the form

$$\sum_{n=0}^{\infty} c_n(x - x_0)^n.$$

The series converges in the intersection of the two intervals of convergence (of q and r) and I. Substituting this series into Equation (2.4) allows us to express the coefficients c_n, for all $n \in \{2, 3, 4, \ldots\}$, in terms of c_0 and c_1, which remain arbitrary.

With $I = [a, b]$, the solutions of Equation (2.1) may be subjected to boundary conditions at a and b. These can take one of the following forms:

$$\begin{aligned} &(i)\ \ y(c) = \xi, &&y'(c) = \eta,\ c \in \{a, b\},\\ &(ii)\ \ y(a) = \xi, &&y(b) = \eta,\\ &(iii)\ \ y'(a) = \xi, &&y'(b) = \eta. \end{aligned}$$

When the boundary conditions are given at the same point c, as in (i), they are often referred to as initial conditions as mentioned earlier. For the purpose of obtaining a unique solution of Equation (2.1), the point c need not, in general, be one of the endpoints of the interval I, and can be any interior point. But in this presentation, as in most physical applications, boundary (or initial) conditions are always imposed at the endpoints of I. The forms (i) to (iii) of the boundary conditions may be generalized by the pair of equations

$$\alpha_1 y(a) + \alpha_2 y'(a) + \alpha_3 y(b) + \alpha_4 y'(b) = \xi, \tag{2.6}$$

$$\beta_1 y(b) + \beta_2 y'(b) + \beta_3 y(a) + \beta_4 y'(a) = \eta, \tag{2.7}$$

where α_i and β_i are constants that satisfy $\sum_{i=1}^{4} |\alpha_i| > 0$ and $\sum_{i=1}^{4} |\beta_i| > 0$, that is, such that not all the α_i or β_i are zeros. The system of Equations (2.1), (2.6), and (2.7) is called a *boundary-value problem.*

The boundary conditions (2.6) and (2.7) are called *homogeneous* if $\xi = \eta = 0$, and *separated* if $\alpha_3 = \alpha_4 = \beta_3 = \beta_4 = 0$. Separated boundary conditions, which have the form

$$\alpha_1 y(a) + \alpha_2 y'(a) = \xi, \qquad \beta_1 y(b) + \beta_2 y'(b) = \eta, \tag{2.8}$$

are of particular significance in this study. Another important pair of homogeneous conditions, which result from a special choice of the coefficients in (2.6) and (2.7), is given by

$$y(a) = y(b), \qquad y'(a) = y'(b). \tag{2.9}$$

Equations (2.9) are called *periodic boundary conditions.* Note that periodic conditions are coupled, not separated.

Definition 2.1

For any two functions $f, g \in C^1$ the determinant

$$W(f,g)(x) = \begin{vmatrix} f(x) & g(x) \\ f'(x) & g'(x) \end{vmatrix} = f(x)g'(x) - g(x)f'(x)$$

is called the *Wronskian* of f and g. The symbol $W(f,g)(x)$ is sometimes abbreviated to $W(x)$.

The Wronskian derives its significance in the study of differential equations from the following lemmas.

Lemma 2.2

If y_1 and y_2 are solutions of the homogeneous equation

$$y'' + q(x)y' + r(x)y = 0, \qquad x \in I, \tag{2.10}$$

where $q \in C(I)$, then either $W(y_1, y_2)(x) = 0$ for all $x \in I$, or $W(y_1, y_2)(x) \neq 0$ for any $x \in I$.

Proof

From Definition 2.1,

$$W' = y_1 y_2'' - y_2 y_1''.$$

Because y_1 and y_2 are solutions of Equation (2.10), we have

$$y_1'' + qy_1' + ry_1 = 0,$$
$$y_2'' + qy_2' + ry_2 = 0.$$

Multiplying the first equation by y_2, the second by y_1, and subtracting, yields

$$y_1 y_2'' - y_2 y_1'' + q(y_1 y_2' - y_2 y_1') = 0$$
$$\Rightarrow \quad W' + qW = 0.$$

Integrating this last equation, we obtain

$$W(x) = c \exp\left(- \int_a^x q(t)dt \right), \qquad x \in I, \tag{2.11}$$

where c is an arbitrary constant. The exponential function does not vanish for any (real or complex) exponent, therefore $W(x) = 0$ if, and only if, $c = 0$. □

Remark 2.3

With $q \in C(I)$, the expression (2.11) implies that both W and W' are continuous.

Lemma 2.4

Any two solutions y_1 and y_2 of Equation (2.10) are linearly independent if, and only if, $W(y_1, y_2)(x) \neq 0$ on I.

Proof

If y_1 and y_2 are linearly dependent, then one of them is a constant multiple of the other, and therefore $W(y_1, y_2)(x) = 0$ on I. Conversely, if $W(y_1, y_2)(x) = 0$ at any point in I, then, by Lemma 2.2, $W(y_1, y_2)(x) \equiv 0$. From the properties of determinants, this implies that the vector functions (y_1, y_1') and (y_2, y_2') are linearly dependent, and hence y_1 and y_2 are linearly dependent. □

Remark 2.5

We used the fact that y_1 and y_2 are solutions of Equation (2.10) only in the second part of the proof, the "only if" part. That is because the Wronskian of two linearly independent functions may vanish at some, but not all, points in I. Consider, for example, x and x^2 on $[-1, 1]$.

Example 2.6

The equation

$$y'' + y = 0 \tag{2.12}$$

has the two linearly independent solutions, $\sin x$ and $\cos x$. Hence the general solution is

$$y(x) = c_1 \cos x + c_2 \sin x.$$

Note that

$$W(\cos x, \sin x) = \cos^2 x + \sin^2 x = 1 \qquad \text{for all } x \in \mathbb{R}.$$

If Equation (2.12) is given on the interval $[0, \pi]$ subject to the initial conditions

$$y(0) = 0, \qquad y'(0) = 1,$$

we obtain the unique solution

$$y(x) = \sin x.$$

But the homogeneous boundary conditions

$$y(0) = 0, \qquad y'(0) = 0,$$

yield the trivial solution $y = 0$, as would be expected.

On the other hand, the boundary conditions

$$y(0) = 0, \qquad y(\pi) = 0,$$

do not give a unique solution because the pair of equations

$$y(0) = c_1 \cos 0 + c_2 \sin 0 = c_1 = 0$$

$$y(\pi) = c_1 \cos \pi + c_2 \sin \pi = -c_1 = 0$$

does not determine the constant c_2. The determinant of the coefficients in this system is

$$\begin{vmatrix} \cos 0 & \sin 0 \\ \cos \pi & \sin \pi \end{vmatrix} = 0.$$

This last example indicates that the boundary conditions (2.6) and (2.7) do not uniquely determine the constants c_1 and c_2 in the general solution in all cases. But the (initial) conditions

$$y(x_0) = \xi, \qquad y'(x_0) = \eta,$$

always give a unique solution, because the determinant of the coefficients in the system of equations

$$c_1 y_1(x_0) + c_2 y_2(x_0) = \xi$$

$$c_1 y_2'(x_0) + c_2 y_2'(x_0) = \eta$$

is given by

$$\begin{vmatrix} y_1(x_0) & y_2(x_0) \\ y_1'(x_0) & y_2'(x_0) \end{vmatrix} = W(y_1, y_2)(x_0),$$

which cannot vanish according to Lemma 2.4.

In general, given a second-order differential equation on $[a, b]$, the separated boundary conditions

$$\alpha_1 y(a) + \alpha_2 y'(a) = \xi,$$

$$\beta_1 y(b) + \beta_2 y'(b) = \eta,$$

imposed on the general solution $y = c_1y_1 + c_2y_2$, yield the system

$$c_1[\alpha_1y_1(a) + \alpha_2y_1'(a)] + c_2[\alpha_1y_2(a) + \alpha_2y_2'(a)] = \xi,$$
$$c_1[\beta_1y_1(b) + \beta_2y_1'(b)] + c_2[\beta_1y_2(b) + \beta_2y_2'(b)] = \eta.$$

Hence the constants c_1 and c_2 are uniquely determined if, and only if,

$$\begin{vmatrix} (\alpha_1y_1 + \alpha_2y_1')(a) & (\alpha_1y_2 + \alpha_2y_2')(a) \\ (\beta_1y_1 + \beta_2y_1')(b) & (\beta_1y_2 + \beta_2y_2')(b) \end{vmatrix} \neq 0. \tag{2.13}$$

EXERCISES

2.1 Find the general solution of each of the following equations.

(a) $y'' - 4y' + 7y = e^x$.

(b) $xy'' - y' = 3x^2$.

(c) $x^2y'' + 3xy' + y = x - 1$.

2.2 Use the transformation $t = \sqrt{x}$ to solve $xy'' + y'/2 - y = 0$.

2.3 Use power series to solve the equation $y'' + 2xy' + 4y = 0$ about the point $x = 0$. Determine the interval of convergence of the solution.

2.4 Solve the initial-value problem

$$y'' + \frac{1}{1-x}(xy' - y) = 0, \qquad -1 < x < 1,$$
$$y(0) = 0, \quad y'(0) = 1.$$

2.5 If φ_1, φ_2, and φ_3 are solutions of Equation (2.4), prove that

$$\begin{vmatrix} \varphi_1 & \varphi_2 & \varphi_3 \\ \varphi_1' & \varphi_2' & \varphi_3' \\ \varphi_1'' & \varphi_2'' & \varphi_3'' \end{vmatrix} = 0.$$

2.6 If y_1 and y_2 are linearly independent solutions of Equation (2.4), show that

$$q = \frac{y_2y_1'' - y_1y_2''}{W(y_1, y_2)}, \qquad r = \frac{y_1'y_2'' - y_2'y_1''}{W(y_1, y_2)}.$$

2.7 For each of the following pair of solutions, determine the corresponding differential equation of the form $a_0y'' + a_1y' + a_2y = 0$.

(a) $e^{-x}\cos 2x$, $e^{-x}\sin 2x$.

(b) x, x^{-1},

(c) 1, $\log x$.

2.2 Zeros of Solutions

It is not necessary, nor is it always possible, to solve a differential equation of the type

$$y'' + q(x)y' + r(x)y = 0, \qquad x \in I, \tag{2.14}$$

explicitly in order to study the properties of its solutions. Under certain conditions, the parameters of the equation and its boundary conditions determine these properties completely. In particular, such qualitative features of a solution as the number and distribution of its zeros, its singular points, its asymptotic behaviour, and its orthogonality properties, are all governed by the coefficients q and r and the given boundary conditions. We can therefore attempt to deduce some of these properties by analysing the effect of these coefficients on the behaviour of the solution. In this section we shall investigate the effect of q and r on the distribution of the zeros of the solutions. Orthogonality is addressed in the next two sections.

In Example 2.6 we found that the two solutions of $y'' + y = 0$ on $\mathbb{R}$ had an infinite sequence of alternating zeros distributed uniformly, given by

$$\cdots < -\pi < -\frac{\pi}{2} < 0 < \frac{\pi}{2} < \pi < \frac{3\pi}{2} < \cdots,$$

where $\{n\pi : n \in \mathbb{Z}\}$ are the zeros of $\sin x$ and $\{\pi/2 + n\pi : n \in \mathbb{Z}\}$ are those of $\cos x$. We shall presently find that this is not completely accidental.

A function $f : I \to \mathbb{C}$ is said to have an *isolated zero* at $x_0 \in I$ if $f(x_0) = 0$ and there is a neighbourhood U of x_0 such that $f(x) \neq 0$ for all $x \in I \cap U \backslash \{x_0\}$.

Lemma 2.7

If y is a nontrivial solution of the homogeneous Equation (2.14), then the zeros of y are isolated in I.

Proof

Suppose $y(x_0) = 0$, where y is a solution of (2.14). If $y'(x_0) = 0$, then y is identically 0 by the uniqueness theorem. If $y'(x_0) \neq 0$ then, because y' is continuous on I, there is a neighbourhood U of x_0 where $y' \neq 0$ on $U \cap I$ (see [1]). Consequently y is either strictly increasing or strictly decreasing on $U \cap I$. □

Theorem 2.8 (Sturm Separation Theorem)

If y_1 and y_2 are linearly independent solutions of the equation

$$y'' + q(x)y' + r(x)y = 0, \qquad x \in I,$$

then the zeros of y_1 are distinct from those of y_2, and the two sequences of zeros alternate; that is, y_1 has exactly one zero between any two successive zeros of y_2, and viceversa.

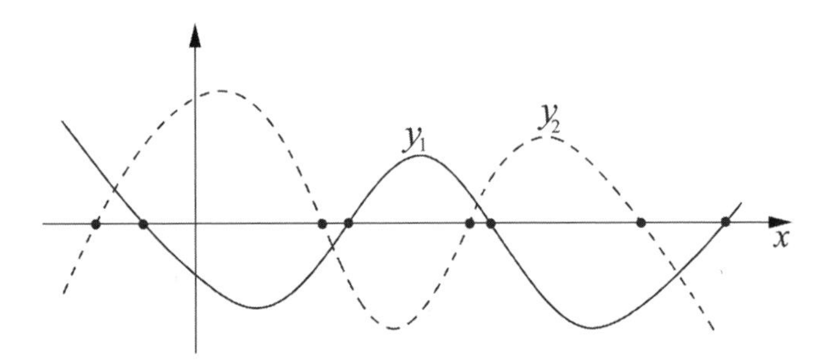

Figure 2.1 The alternating zeros of y_1 and y_2.

Proof

Because y_1 and y_2 are linearly independent, their Wronskian

$$W(y_1, y_2)(x) = y_1(x)y_2'(x) - y_2(x)y_1'(x)$$

does not vanish, and has therefore one sign on I (Lemma 2.4). Note first that y_1 and y_2 cannot have a common zero, otherwise W would vanish there. Suppose x_1 and x_2 are two successive zeros of y_2. Then

$$W(x_1) = y_1(x_1)y_2'(x_1) \neq 0,$$
$$W(x_2) = y_1(x_2)y_2'(x_2) \neq 0,$$

and the numbers $y_1(x_1)$, $y_1(x_2)$, $y_2'(x_1)$, and $y_2'(x_2)$ are all nonzero. Because y_2' is continuous on I, x_1 has a neighborhood U_1 where the sign of y_2' does not change, and similarly there is a neighborhood U_2 of x_2 where y_2' does not change sign. But the signs of y_2' in $U_1 \cap I$ and $U_2 \cap I$ cannot be the same, for if y_2 is increasing on one then it has to be decreasing on the other. For $W(x)$ to have a constant sign on I, $y_1(x_1)$ and $y_1(x_2)$ must therefore have opposite signs; hence y_1, being continuous, has at least one zero between x_1 and x_2. There cannot be more than one such zero, for if x_3 and x_4 are two zeros of y_1 which lie between x_1 and x_2, we can use the same argument to conclude that y_2 vanishes between x_3 and x_4. But this contradicts the assumption that x_1 and x_2 are consecutive zeros of y_2. □

Corollary 2.9

If two solutions of $y'' + q(x)y' + r(x)y = 0$ have a common zero in I, then they are linearly dependent.

In order to study the distribution of zeros of Equation (2.14), it would be much more convenient if we could get rid of the middle term qy' by transforming the equation to

$$u'' + \rho(x)u = 0. \tag{2.15}$$

To that end we set

$$y(x) = u(x)v(x),$$

so that

$$\begin{aligned} y'(x) &= u'(x)v(x) + u(x)v'(x), \\ y''(x) &= u''(x)v(x) + 2u'(x)v'(x) + u(x)v''(x). \end{aligned}$$

Substituting into Equation (2.14) yields

$$vu'' + (2v' + qv)u' + (v'' + qv' + rv)u = 0.$$

Thus we obtain (2.15) by choosing $2v' + qv = 0$, which implies

$$\begin{aligned} v(x) &= \exp\left(-\frac{1}{2}\int_a^x q(t)dt\right), \\ \rho(x) &= r(x) - \frac{1}{4}q^2(x) - \frac{1}{2}q'(x). \end{aligned} \tag{2.16}$$

The exponential function v never vanishes on $\mathbb{R}$, thus the zeros of u coincide with those of y, and we may, for the purpose of investigating the distribution of zeros of Equation (2.14), confine our attention to Equation (2.15).

Theorem 2.10 (Sturm Comparison Theorem)

Let φ and ψ be nontrivial solutions of the equations

$$\begin{aligned} y'' + r_1(x)y &= 0, \\ u'' + r_2(x)u &= 0, \qquad x \in I, \end{aligned}$$

respectively, and suppose $r_1(x) \geq r_2(x)$ for all $x \in I$. Then φ has at least one zero between every two consecutive zeros of ψ, unless $r_1(x) \equiv r_2(x)$ and φ is a constant multiple of ψ.

Proof

Let x_1 and x_2 be any two consecutive zeros of ψ on I, and suppose that φ has no zero in the open interval (x_1, x_2). Assume that both φ and ψ are positive on (x_1, x_2), otherwise change the sign of the negative function. Since φ' and ψ' are continuous, it follows that $\psi'(x_1) \geq 0$ and $\psi'(x_2) \leq 0$, and therefore the Wronskian of φ and ψ satisfies

$$W(x_1) = \varphi(x_1)\psi'(x_1) \geq 0, \qquad W(x_2) = \varphi(x_2)\psi'(x_2) \leq 0. \tag{2.17}$$

But because

$$\begin{aligned} W'(x) &= \varphi(x)\psi''(x) - \varphi''(x)\psi(x) \\ &= [r_1(x) - r_2(x)]\varphi(x)\psi(x) \geq 0 \qquad \text{for all } x \in (x_1, x_2), \end{aligned}$$

W is an increasing function on (x_1, x_2). This contradicts Equation (2.17) unless $r_1(x) - r_2(x) \equiv 0$ and $W(x) \equiv 0$, in which case φ and ψ are linearly dependent (by Lemma 2.4). □

Corollary 2.11

Let φ be a nontrivial solution of $y'' + r(x)y = 0$ on I. If $r(x) \leq 0$ then φ has at most one zero on I.

Proof

If the solution φ has two zeros on I, say x_1 and x_2, then, by Theorem 2.10, the solution $\psi(x) \equiv 1$ of $u'' = 0$ must vanish on (x_1, x_2), which is impossible. □

Example 2.12

(i) Any nontrivial solution of $y'' = 0$ on $\mathbb{R}$ is a special case of

$$\varphi(x) = c_1 x + c_2,$$

which is represented by a straight line, and has at most one zero.

(ii) The equation $y'' - y = 0$ has the general solution

$$\varphi(x) = c_1 e^x + c_2 e^{-x}, \qquad x \in \mathbb{R}.$$

If c_1 and c_2 are not both zero, then $\varphi(x) \neq 0$ for any $x \in \mathbb{R}$ unless $c_2 = -c_1$, in which case φ has one zero at $x = 0$.

(iii) The solution of $y'' + y = 0$ is given by

$$\varphi(x) = c_1 \cos x + c_2 \sin x = a \sin(x - b),$$

where $a = \sqrt{c_1^2 + c_2^2}$ and $b = -\arctan(c_1/c_2)$. If $a \neq 0$, φ has an infinite number of zeros given by $x_n = b + n\pi,\ n \in \mathbb{Z}$.

A nontrivial solution of

$$y'' + r(x)y = 0, \qquad x \in I, \tag{2.18}$$

is called *oscillatory* if it has an infinite number of zeros, as in Example 2.12(iii). According to Theorem 2.10, whether this equation has oscillatory solutions depends on the function r. If $r(x) \leq 0$, the solutions cannot oscillate by Corollary 2.11; but if

$$r(x) > k^2 > 0, \qquad x \in I,$$

for some positive constant k, then any solution of (2.18) on I has an infinite number of zeros distributed between the zeros of any solution of $y'' + k^2 y = 0$, such as $a \sin k(x - b)$, which are given by

$$x_n = b + \frac{n\pi}{k}.$$

Thus every subinterval of I of length π/k has at least one zero of Equation (2.18), and as k increases we would expect the number of zeros to increase. This, of course, is clearly the case when r is constant.

From the Sturm separation theorem we also conclude that, if the interval I is infinite and one solution of the equation

$$y'' + q(x)y' + r(x)y = 0$$

oscillates, then all its solutions oscillate.

Example 2.13

The equation

$$y'' + \frac{1}{x}y' + \left(1 - \frac{\nu^2}{x^2}\right)y = 0, \qquad 0 < x < \infty, \tag{2.19}$$

is known as Bessel's equation of order ν, and is the subject of Chapter 5. Using formula (2.16) to define $u = \sqrt{x}y$, Bessel's equation under this transformation takes the form

$$u'' + \left(1 + \frac{1 - 4\nu^2}{4x^2}\right)u = 0. \tag{2.20}$$

By comparing Equations (2.20) and $u'' + u = 0$, we see that

$$r(x) = 1 + \frac{1 - 4\nu^2}{4x^2} \begin{cases} \geq 1 & \text{if } 0 \leq \nu \leq 1/2 \\ < 1 & \text{if } \nu > 1/2. \end{cases}$$

Applying Theorem 2.10 we therefore conclude:

(i) If $0 \leq \nu \leq 1/2$ then, in every subinterval of $(0, \infty)$ of length π, any solution of Bessel's equation has at least one zero.

(ii) If $\nu > 1/2$ then, in every subinterval of $(0, \infty)$ of length π, any nontrivial solution of Bessel's equation has at most one zero.

(iii) If $\nu = 1/2$, the distance between successive zeros of any nontrivial solution of Bessel's equation is exactly π.

EXERCISES

2.8 Prove that any nontrivial solution of $y'' + r(x)y = 0$ on a finite interval has at most a finite number of zeros.

2.9 Prove that any nontrivial solution of the equation

$$y'' + \frac{k}{x^2}y = 0$$

on $(0, \infty)$ is oscillatory if, and only if, $k > 1/4$. Hint: Use the substitution $x = e^t$.

2.10 Use the result of Exercise 2.9 to conclude that any nontrivial solution of the equation $y'' + r(x)y = 0$ on $(0, \infty)$ has an infinite number of zeros if $r(x) \geq k/x^2$ for some $k > 1/4$, and only a finite number if $r(x) < 1/4x^2$.

2.11 Let φ be a nontrivial solution of $y'' + r(x)y = 0$ on $(0, \infty)$, where $r(x) > 0$. If $\varphi(x) > 0$ on $(0, a)$ for some positive number a, and if there is a point $x_0 \in (0, a)$ where $\varphi'(x_0) < 0$, prove that φ vanishes at some point $x_1 > x_0$.

2.12 Determine which equations have oscillatory solutions on $(0, \infty)$:

(a) $y'' + (\sin^2 x + 1)y = 0$.

(b) $y'' - x^2 y = 0$.

(c) $y'' + \dfrac{1}{x}y = 0$.

2.13 Find the general solution of Bessel's equation of order $1/2$, and determine the zeros of each independent solution.

2.14 If $\lim_{x\to\infty} f(x) = 0$, prove that the solutions of $y'' + (1+f(x))y = 0$ are oscillatory.

2.15 Prove that any solution of Airy's equation $y''+xy = 0$ has an infinite number of zeros on $(0,\infty)$, and at most one zero on $(-\infty, 0)$.

2.3 Self-Adjoint Differential Operator

Going back to the general form of the linear second-order differential Equation (2.1), in slightly modified notation,

$$p(x)y'' + q(x)y' + r(x)y = 0, \tag{2.21}$$

we now wish to investigate the orthogonality properties of its solutions. This naturally means we should look for the C^2 solutions of (2.21) which lie in $\mathcal{L}^2$. Equation (2.21) can be written in the form

$$Ly = 0,$$

where

$$L = p(x)\frac{d^2}{dx^2} + q(x)\frac{d}{dx} + r(x), \tag{2.22}$$

is a linear differential operator of second order, and y lies in $\mathcal{L}^2(I) \cap C^2(I)$, which is a linear vector space, being the intersection of two such spaces with the same operations.

To motivate the discussion, it would help at this point to recall some of the notions of linear algebra. A *linear operator* in a vector space X is a mapping

$$A : X \to X$$

which satisfies

$$A(a\mathbf{x} + b\mathbf{y}) = aA\mathbf{x} + bA\mathbf{y} \quad \text{for all } a, b \in \mathbb{F}, \qquad \mathbf{x}, \mathbf{y} \in X.$$

If X is an inner product space, the *adjoint* of A, if it exists, is the operator A' which satisfies

$$\langle A\mathbf{x}, \mathbf{y}\rangle = \langle \mathbf{x}, A'\mathbf{y}\rangle \quad \text{for all } \mathbf{x}, \mathbf{y} \in X.$$

If $A' = A$, then A is said to be *self-adjoint.*

If X is a finite-dimensional inner product space, such as $\mathbb{C}^n$ over $\mathbb{C}$, we know that any linear operator is represented, with respect to the orthonormal basis $\{\mathbf{e}_i : 1 \le i \le n\}$, by the matrix

$$A = \begin{pmatrix} a_{11} & \cdots & a_{1n} \\ \vdots & & \vdots \\ a_{n1} & \cdots & a_{nn} \end{pmatrix} = (a_{ij}).$$

Its adjoint is given by

$$A' = \begin{pmatrix} \bar{a}_{11} & \cdots & \bar{a}_{n1} \\ \vdots & & \vdots \\ \bar{a}_{1n} & \cdots & \bar{a}_{nn} \end{pmatrix} = (\bar{a}_{ji}) = \bar{A}^T,$$

where $\bar{A}$ is the *complex conjugate of* A,

$$\bar{A} = \begin{pmatrix} \bar{a}_{11} & \cdots & \bar{a}_{1n} \\ \vdots & & \vdots \\ \bar{a}_{n1} & \cdots & \bar{a}_{nn} \end{pmatrix},$$

and A^T is its *transpose*,

$$A^T = \begin{pmatrix} a_{11} & \cdots & a_{n1} \\ \vdots & & \vdots \\ a_{1n} & \cdots & a_{nn} \end{pmatrix}.$$

A complex number a is called an *eigenvalue* of A if there is a nonzero vector $\mathbf{x} \in X$ such that

$$A\mathbf{x} = a\mathbf{x},$$

and $\mathbf{x}$ in this case is called an *eigenvector* of the operator A associated with the eigenvalue a. From linear algebra (see, for example, [11]) we know that, if A is a self-adjoint (or *Hermitian*) matrix, then

(i) The eigenvalues of A are all real numbers.

(ii) The eigenvectors of A corresponding to distinct eigenvalues are orthogonal.

(iii) The eigenvectors of A form a basis of X.

In order to extend these results to the space $\mathcal{L}^2$, our first task is to obtain the form of the adjoint of the operator

$$L : \mathcal{L}^2(I) \cap C^2(I) \to \mathcal{L}^2(I)$$

defined by Equation (2.22), where we assume, to start with, that the coefficients p, q, and r are C^2 functions on I. Note that $C^2(I) \cap \mathcal{L}^2(I) = C^2(I)$ when I is a closed and bounded interval. Denoting the adjoint of L by L', we have, by definition of L',

$$\langle Lf, g \rangle = \langle f, L'g \rangle \qquad \text{for all } f, g \in C^2(I) \cap \mathcal{L}^2(I). \tag{2.23}$$

We set $I = (a, b)$, where the interval I may be finite or infinite, and use integration by parts to manipulate the left-hand side of (2.23) in order to shift the differential operator from f to g. Thus

$$\begin{aligned}
\langle Lf, g\rangle &= \int_a^b (pf'' + qf' + rf)\bar{g}dx \\
&= pf'\bar{g}\big|_a^b - \int_a^b f'(p\bar{g})'dx + qf\bar{g}\big|_a^b - \int_a^b f(q\bar{g})'dx + \int_a^b fr\bar{g}dx \\
&= [pf'\bar{g} - f(p\bar{g})']\big|_a^b + \int_a^b f(p\bar{g})''dx + qf\bar{g}\big|_a^b - \int_a^b f(q\bar{g})'dx + \int_a^b fr\bar{g}dx \\
&= \langle f, (\bar{p}g)'' - (\bar{q}g)' + \bar{r}g\rangle + [p(f'\bar{g} - f\bar{g}') + (q - p')f\bar{g}]\big|_a^b,
\end{aligned}$$

where the integrals are considered improper if (a, b) is infinite or any of the integrands is unbounded at a or b. Note that the right-hand side of the above equation is well defined if $p \in C^2(a, b)$, $q \in C^1(a, b)$, and $r \in C(a, b)$. The last term, of course, is to be interpreted as the difference between the limits at a and b. We therefore have, for all $f, g \in \mathcal{L}^2(I) \cap C^2(I)$,

$$\langle Lf, g\rangle = \langle f, L^*g\rangle + [p(f'\bar{g} - f\bar{g}') + (q - p')f\bar{g}]\big|_a^b, \tag{2.24}$$

where

$$\begin{aligned}
L^*g &= (\bar{p}g)'' - (\bar{q}g)' + \bar{r}g \\
&= \bar{p}g'' + (2\bar{p}' - \bar{q})g' + (\bar{p}'' - \bar{q}' + \bar{r})g.
\end{aligned}$$

The operator

$$L^* = \bar{p}\frac{d^2}{dx^2} + (2\bar{p}' - \bar{q})\frac{d}{dx} + (\bar{p}'' - \bar{q}' + \bar{r})$$

is called the *formal adjoint* of L. L is said to be *formally self-adjoint* when $L^* = L$, that is, when

$$\bar{p} = p, \qquad 2\bar{p}' - \bar{q} = q, \qquad \bar{p}'' - \bar{q}' + \bar{r} = r.$$

These three equations are satisfied if, and only if, the functions p, q, and r are real and $q = p'$. In that case

$$\begin{aligned}
Lf &= pf'' + p'f' + rf \\
&= (pf')' + rf.
\end{aligned}$$

Thus, when L is formally self-adjoint, it has the form

$$L = \frac{d}{dx}\left(p\frac{d}{dx}\right) + r,$$

and Equation (2.24) is reduced to

$$\langle Lf, g\rangle = \langle f, Lg\rangle + p(f'\bar{g} - f\bar{g}')\big|_a^b. \tag{2.25}$$

Comparing Equations (2.23) and (2.25) we now see that the formally self-adjoint operator L is self-adjoint if

$$p(f'\bar{g} - f\bar{g}')\big|_a^b = 0 \quad \text{for all} f, g \in C^2(I) \cap \mathcal{L}^2(I). \tag{2.26}$$

It is worth noting at this point that, when $q = p'$, the term $\bar{p}'' - \bar{q}'$ in the expression for L^* drops out, hence the continuity of p'' and q' is no longer needed.

We are interested in the eigenvalue problem for the operator $-L$, that is, solutions of the equation

$$Lu + \lambda u = 0. \tag{2.27}$$

When $u = 0$ this equation, of course, is satisfied for every value of λ. When $u \neq 0$, it may be satisfied for certain values of λ. These are the eigenvalues of $-L$. Any function $u \neq 0$ in $C^2 \cap \mathcal{L}^2$ which satisfies Equation (2.27) for some complex number λ is an eigenfunction of $-L$ corresponding to the eigenvalue λ. We also refer to the eigenvalues and eigenfunctions of $-L$ as eigenvalues and eigenfunctions of Equation (2.27). Because this equation is homogeneous, the eigenfunctions of $-L$ are determined up to a multiplicative constant. When appropriate boundary conditions are added to Equation (2.27), the resulting system is called a *Sturm–Liouville eigenvalue problem*, which is discussed in the next section. Clearly, $-L$ is (formally) self-adjoint if, and only if, L is (formally) self-adjoint. The reason we look for the eigenvalues of $-L$ rather than L is that, as it turns out, L has negative eigenvalues when p is positive (see Example 2.16 below). The following theorem summarizes the results we have obtained thus far, as it generalizes properties (i) and (ii) above from finite-dimensional space to the infinite-dimensional space $\mathcal{L}^2 \cap C^2$.

Theorem 2.14

Let $L : \mathcal{L}^2(a,b) \cap C^2(a,b) \to \mathcal{L}^2(a,b)$ be a linear differential operator of second order defined by

$$Lu = p(x)u'' + q(x)u' + r(x)u, \quad x \in (a,b),$$

where $p \in C^2(a,b)$, $q \in C^1(a,b)$, and $r \in C(a,b)$. Then

(i) L is formally self-adjoint, that is, $L^* = L$, if the coefficients p, q, and r are real and $q \equiv p'$.

(ii) L is self-adjoint, that is, $L' = L$, if L is formally self-adjoint and Equation (2.26) is satisfied.

(iii) If L is self-adjoint, then the eigenvalues of the equation

$$Lu + \lambda u = 0$$

are all real and any pair of eigenfunctions associated with distinct eigenvalues are orthogonal in $\mathcal{L}^2(a,b)$.

Proof

We have already proved (i) and (ii). To prove (iii), suppose $\lambda \in \mathbb{C}$ is an eigenvalue of $-L$. Then there is a function $f \in \mathcal{L}^2(a,b) \cap C^2(a,b)$, $f \neq 0$, such that

$$\begin{aligned} &Lf + \lambda f = 0 \\ \Rightarrow\ &\lambda \|f\|^2 = \langle \lambda f, f \rangle = -\langle Lf, f \rangle . \end{aligned}$$

Because L is self-adjoint,

$$-\langle Lf, f \rangle = -\langle f, Lf \rangle = \langle f, \lambda f \rangle = \bar{\lambda} \|f\|^2 .$$

Hence $\bar{\lambda} \|f\|^2 = \lambda \|f\|^2$. Because $\|f\| \neq 0$, $\bar{\lambda} = \lambda$.

If μ is another eigenvalue of $-L$ associated with the eigenfunction $g \in \mathcal{L}^2(a,b) \cap C^2(a,b)$, then

$$\begin{aligned} \lambda \langle f, g \rangle = -\langle Lf, g \rangle = -\langle f, Lg \rangle &= \mu \langle f, g \rangle \\ (\lambda - \mu) \langle f, g \rangle &= 0 \\ \lambda \neq \mu \ \Rightarrow\ \langle f, g \rangle &= 0. \end{aligned}$$

□

Remark 2.15

As noted earlier, when $q = p'$ in the expression for L, the conclusions of this theorem are valid under the weaker requirement that p' be continuous.

Example 2.16

The operator $-(d^2/dx^2)$ is formally self-adjoint with $p = -1$ and $r = 0$. To determine its eigenfunctions in $C^2(0,\pi)$, we have to solve the equation

$$u'' + \lambda u = 0.$$

Let us first consider the case where $\lambda > 0$. The general solution of the equation is given by

$$u(x) = c_1 \cos \sqrt{\lambda} x + c_2 \sin \sqrt{\lambda} x. \tag{2.28}$$

Under the boundary conditions

$$u(0) = u(\pi) = 0$$

Equation (2.26) is satisfied, so $-(d^2/dx^2)$ is, in fact, self-adjoint. Applying the boundary conditions to (2.28), we obtain

$$u(0) = c_1 = 0$$
$$u(\pi) = c_2 \sin\sqrt{\lambda}\pi = 0 \Rightarrow \sqrt{\lambda}\pi = n\pi \Rightarrow \lambda = n^2, \quad n \in \mathbb{N}.$$

Thus the eigenvalues of $-(d^2/dx^2)$ are given by the sequence

$$(n^2 : n \in \mathbb{N}) = (1, 4, 9, \ldots),$$

and the corresponding eigenfunctions are

$$(\sin nx : n \in \mathbb{N}) = (\sin x, \sin 2x, \sin 3x, \ldots).$$

Observe that we chose $c_2 = 1$ for the sake of simplicity, since the boundary conditions as well as the eigenvalue equation are all homogeneous. We could also divide each eigenfunction by its norm $\|\sin nx\| = \sqrt{\int_0^\pi \sin^2 x \, dx} = \sqrt{\pi/2}$ to obtain the normalized eigenfunctions

$$\left(\sqrt{2/\pi}\sin nx : n \in \mathbb{N}\right).$$

If $\lambda = 0$, the solution of the differential equation is given by $c_1 x + c_2$, and if $\lambda < 0$ it is given by $c_1 \cosh\sqrt{-\lambda}x + c_2 \sinh\sqrt{-\lambda}x$. In either case the application of the boundary conditions at $x = 0$ and $x = \pi$ leads to the conclusion that $c_1 = c_2 = 0$. But the trivial solution is not admissible as an eigenfunction, therefore we do not have any eigenvalues in the interval $(-\infty, 0]$. The eigenvalues $\lambda_n = n^2$ are real numbers, in accordance with Theorem 2.14, and the eigenfunctions $u_n(x) = \sin nx$ are orthogonal in $\mathcal{L}^2(0, \pi)$ because, for all $n \neq m$,

$$\int_0^\pi \sin nx \sin mx dx = \frac{1}{2}\left[\frac{\sin(n-m)x}{n-m} - \frac{\sin(n+m)x}{n+m}\right]\Bigg|_0^\pi = 0.$$

Sometimes it is not possible to determine the eigenvalues of a system exactly, as the next example shows.

Example 2.17

Consider the same equation

$$u'' + \lambda u = 0$$

on the interval $(0, l)$, under the separated boundary conditions

$$u(0) = 0, \qquad hu(l) + u'(l) = 0,$$

where h is a positive constant.

It is straightforward to check that this system has no eigenvalues in $(-\infty, 0]$. When $\lambda > 0$ the general solution is

$$u(x) = c_1 \cos \sqrt{\lambda} x + c_2 \sin \sqrt{\lambda} x.$$

The first boundary condition implies $c_1 = 0$, and the second yields

$$c_2(h \sin \sqrt{\lambda} l + \sqrt{\lambda} \cos \sqrt{\lambda} l) = 0.$$

Because c_2 cannot be 0, otherwise we do not get an eigenfunction, we must have

$$h \sin \sqrt{\lambda} l + \sqrt{\lambda} \cos \sqrt{\lambda} l = 0.$$

Because $\sin \sqrt{\lambda} l$ and $\cos \sqrt{\lambda} l$ cannot both be 0, it follows that neither of them is. Therefore we can divide by $\cos \sqrt{\lambda} l$ to obtain

$$\tan \sqrt{\lambda} l = -\frac{\sqrt{\lambda}}{h}.$$

Setting $\alpha = \sqrt{\lambda} l$, we see that α is determined by the equation

$$\tan \alpha = -\frac{\alpha}{hl}.$$

This is a transcendental equation which is solved graphically by the points of intersection of the graphs of

$$y = \tan \alpha \quad \text{and} \quad y = -\frac{\alpha}{hl},$$

as shown by the sequence (α_n) in Figure 2.2.

The eigenvalues and eigenfunctions of the problem are therefore given by

$$\begin{aligned} \lambda_n &= \left(\frac{\alpha_n}{l}\right)^2, \\ u_n(x) &= \sin(\alpha_n x / l), \qquad 0 \le x \le l, \qquad n \in \mathbb{N}, \end{aligned}$$

and the conclusions of Theorem 2.14(iii) are clearly satisfied.

The following example shows that, if $p' \neq q$ on I, the operator L in Theorem 2.14 may be transformed to a formally self-adjoint operator when multiplied by a suitable function.

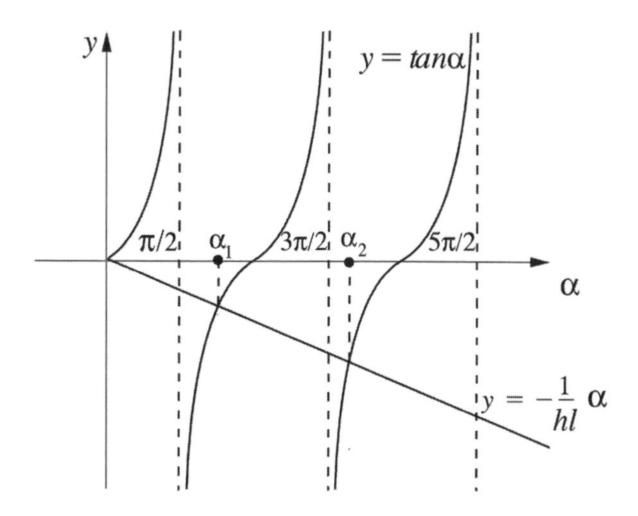

Figure 2.2

Example 2.18

Let

$$L = p(x)\frac{d^2}{dx^2} + q(x)\frac{d}{dx} + r(x), \qquad x \in I = [a, b],$$

where $p \in C^2(I)$ does not vanish on I, $q \in C^1(I)$, and $r \in C(I)$. We may assume, without loss of generality, that $p(x) > 0$ for all $x \in I$. If $q \neq p'$, we can multiply L by a function ρ and define the operator

$$\tilde{L} = \rho L = \rho p \frac{d^2}{dx^2} + \rho q \frac{d}{dx} + \rho r.$$

$\tilde{L}$ will be formally self-adjoint if

$$\rho q = (\rho p)' = \rho' p + \rho p'.$$

This is a first-order differential equation in ρ, whose solution is

$$\rho(x) = \frac{c}{p(x)} \exp\left(\int_a^x \frac{q(t)}{p(t)} dt\right), \tag{2.29}$$

where c is a constant. Note that ρ is a C^2 function which is strictly positive on I. It reduces to a (nonzero) constant when $q = p'$, that is, when L is formally self-adjoint.

The result of Example 2.18 allows us to generalize part (iii) of Theorem 2.14 to differential operators which are not formally self-adjoint. If $Lu = pu'' + qu' + ru$, where $p > 0$ and $q \neq p'$, the eigenvalue equation

$$Lu + \lambda u = 0$$

can be multiplied by the positive function ρ defined by (2.29) to give the equivalent equation

$$\rho Lu + \lambda \rho u = 0, \tag{2.30}$$

where ρL is now formally self-adjoint. With $\rho > 0$, Equation (2.26) is equivalent to

$$\rho p(f'\bar{g} - f\bar{g}')|_a^b = 0.$$

This makes the operator ρL self-adjoint. If $u \in \mathcal{L}_\rho^2$ is an eigenfunction of L associated with the eigenvalue λ, we can write

$$\begin{aligned}\lambda \|u\|_\rho^2 &= \langle \lambda \rho u, u \rangle \\ &= \langle -\rho Lu, u \rangle \\ &= \langle u, -\rho Lu \rangle \\ &= \langle u, \lambda \rho u \rangle \\ &= \bar{\lambda} \|u\|_\rho^2 ,\end{aligned}$$

which implies that λ is a real number. Furthermore, if $v \in \mathcal{L}_\rho^2$ is another eigenfunction of L associated with the eigenvalue μ, then

$$\begin{aligned}(\lambda - \mu) \langle u, v \rangle_\rho &= \lambda \langle \rho u, v \rangle - \mu \langle \rho u, v \rangle \\ &= \langle \lambda \rho u, v \rangle - \langle u, \mu \rho v \rangle \\ &= \langle -\rho Lu, v \rangle - \langle u, -\rho Lv \rangle = 0\end{aligned}$$

because ρL is self-adjoint. Thus, if $\lambda \neq \mu$, then u is orthogonal to v in $\mathcal{L}_\rho^2$.

We have therefore proved the following extension of Theorem 2.14(iii):

Corollary 2.19

If $L : \mathcal{L}^2(a, b) \cap C^2(a, b) \to \mathcal{L}^2(a, b)$ is a self-adjoint linear operator and ρ is a positive and continuous function on $[a, b]$, then the eigenvalues of the equation

$$Lu + \lambda \rho u = 0$$

are all real and any pair of eigenfunctions associated with distinct eigenvalues are orthogonal in $\mathcal{L}_\rho^2(a, b)$.

Remark 2.20

1. In this corollary the eigenvalues and eigenfunctions of the equation $Lu + \lambda \rho u = 0$ are, in fact, the eigenvalues and eigenfunctions of the operator $-\rho^{-1}L$.

2. Suppose the interval (a, b) is finite. Since the weight function ρ is continuous and positive on $[a, b]$, its minimum value α and its maximum value β satisfy

$$0 < \alpha \leq \rho(x) \leq \beta < \infty.$$

This implies

$$\sqrt{\alpha} \, \|u\| \leq \|u\|_\rho \leq \sqrt{\beta} \, \|u\| \, ,$$

and therefore $\|u\|_\rho < \infty$ if, and only if, $\|u\| < \infty$. The two norms are said to be equivalent, and the two spaces $\mathcal{L}^2_\rho(a, b)$ and $\mathcal{L}^2(a, b)$ clearly contain the same functions, although they have different inner products.

3. Nothing in the proof of this corollary requires L to be a second-order differential operator. In fact, the result is true for any self-adjoint linear operator on an inner product space.

Example 2.21

Find the eigenfunctions and eigenvalues of the boundary-value problem

$$x^2 y'' + xy' + \lambda y = 0, \qquad 1 < x < b, \tag{2.31}$$

$$y(1) = y(b) = 0. \tag{2.32}$$

Solution

Equation (2.31) is a Cauchy–Euler equation whose solutions have the form x^m. Substituting into the equation leads to

$$m(m-1) + m + \lambda = 0$$
$$\Rightarrow m = \pm i\sqrt{\lambda}.$$

Assuming $\lambda > 0$, we have

$$x^{i\sqrt{\lambda}} = e^{i\sqrt{\lambda} \log x} = \cos(\sqrt{\lambda} \log x) + i \sin(\sqrt{\lambda} \log x),$$

and the general solution of Equation (2.31) is given by

$$y(x) = c_1 \cos(\sqrt{\lambda} \log x) + c_2 \sin(\sqrt{\lambda} \log x).$$

Applying the boundary conditions (2.32),

$$y(1) = c_1 = 0, \qquad y(b) = c_2 \sin(\sqrt{\lambda} \log b) = 0.$$

Because we cannot have both constants c_1 and c_2 vanish, this implies

$$\sin(\sqrt{\lambda} \log b) = 0$$
$$\Rightarrow \sqrt{\lambda} \log b = n\pi, \qquad n \in \mathbb{N}.$$

Thus the eigenvalues of the boundary-value problem are given by the sequence of positive real numbers

$$\lambda_n = \left(\frac{n\pi}{\log b} \right)^2, \qquad n \in \mathbb{N},$$

and the corresponding sequence of eigenfunctions is

$$y_n(x) = \sin\left(\frac{n\pi}{\log b} \log x \right).$$

Observe here that the differential operator

$$x^2 \frac{d^2}{dx^2} + x \frac{d}{dx}$$

is not formally self-adjoint, but it becomes so after multiplication by the weight function

$$\rho(x) = \frac{1}{x^2} \exp\left(\int_1^x \frac{1}{t} dt \right) = \frac{1}{x}.$$

The resulting operator is then

$$L = x \frac{d^2}{dx^2} + \frac{d}{dx} = \frac{d}{dx} \left(x \frac{d}{dx} \right).$$

The eigenfunctions $(y_n : n \in \mathbb{N})$ of this problem are in fact eigenfunctions of $-\rho^{-1} L = -xL$, which are orthogonal in $\mathcal{L}^2_\rho(1, b)$, not $\mathcal{L}^2(1, b)$. Indeed,

$$\begin{aligned} \langle y_m, y_n \rangle_\rho &= \int_1^b \sin\left(\frac{m\pi}{\log b} \log x \right) \sin\left(\frac{n\pi}{\log b} \log x \right) \frac{1}{x} dx \\ &= \frac{\log b}{\pi} \int_0^\pi \sin m\xi \sin n\xi d\xi \\ &= 0 \quad \text{for all } m \neq n. \end{aligned}$$

We leave it as an exercise to show that this problem has no eigenvalues in the interval $(-\infty, 0]$.

EXERCISES

2.16 Given $L = \frac{d}{dx}\left(p\frac{d}{dx}\right) + r$, prove the Lagrange identity

$$uLv - vLu = [p(uv' - vu')]'.$$

The integral of this identity,

$$\int_a^b (uLv - vLu)dx = [p(uv' - vu')]\big|_a^b,$$

is known as Green's formula.

2.17 Determine the eigenfunctions and eigenvalues of the differential operators

(a) $\frac{d^2}{dx^2} : C^2(0,\infty) \to C(0,\infty)$.

(b) $\frac{d^2}{dx^2} : \mathcal{L}^2(0,\infty) \cap C^2(0,\infty) \to \mathcal{L}^2(0,\infty)$.

2.18 Prove that the differential operator $-(d^2/dx^2) : \mathcal{L}^2(0,\pi) \cap C^2(0,\pi) \to \mathcal{L}^2(0,\pi)$ is self-adjoint under the boundary conditions $u(0) = u'(\pi) = 0$. Verify that its eigenvalues and eigenfunctions satisfy Theorem 2.14(iii).

2.19 Put each of the following differential operators in the form $p(d^2/dx^2) + p'(d/dx) + r$, with $p > 0$, by multiplying by a suitable weight function:

(a) $x^2\frac{d^2}{dx^2}, \quad x > 0,$

(b) $\frac{d^2}{dx^2} - x\frac{d}{dx}, \quad x \in \mathbb{R},$

(c) $\frac{d^2}{dx^2} - x^2\frac{d}{dx}, \quad x > 0,$

(d) $x^2\frac{d^2}{dx^2} + x\frac{d}{dx} + (x^2 - \lambda), \quad x > 0.$

2.20 Determine the eigenvalues and eigenfunctions of $u'' + u = 0$ on $(0, l)$ subject to the boundary conditions $u(0) = 0$, $hu(l) + u'(l) = 0$, where $h \leq 0$.

2.21 Put the eigenvalue equation $u'' + 2u' + \lambda u = 0$ in the standard form $Lu + \lambda\rho u = 0$, where L is self-adjoint, then find its eigenvalues and eigenfunctions on $[0,1]$ subject to the boundary conditions $u(0) = u(1) = 0$. Verify that the result agrees with Corollary 2.19.

2.22 Determine the eigenfunctions and eigenvalues of the equation $x^2u'' + \lambda u = 0$ on $[1, e]$, subject to the boundary conditions $u(1) = u(e) = 0$.

2.4 The Sturm–Liouville Problem

In Section 2.3 we saw how a linear differential equation may be posed as an eigenvalue problem for a differential operator in an infinite-dimensional vector space. The notion of self-adjointness was defined by drawing on the analogy with linear transformations in finite-dimensional space, and we saw how the first two properties of a self-adjoint matrix, namely that its eigenvalues are real and its eigenvectors orthogonal, translated (almost word for word) into Theorem 2.14. The third property, that its eigenvectors span the space, is in fact an assertion that the matrix has eigenvectors as well as a statement on their number. This section is concerned with deriving the corresponding result for differential operators, which is expressed by Theorems 2.26 and 2.28.

Let L be a formally self-adjoint operator of the form

$$L = \frac{d}{dx}\left(p(x)\frac{d}{dx}\right) + r(x). \tag{2.33}$$

The eigenvalue equation

$$Lu + \lambda\rho(x)u = 0, \qquad x \in (a, b), \tag{2.34}$$

subject to the separated homogeneous boundary conditions

$$\alpha_1 u(a) + \alpha_2 u'(a) = 0, \qquad |\alpha_1| + |\alpha_2| > 0, \tag{2.35}$$

$$\beta_1 u(b) + \beta_2 u'(b) = 0, \qquad |\beta_1| + |\beta_2| > 0, \tag{2.36}$$

where α_i and β_i are real constants, is called a *Sturm–Liouville eigenvalue problem*, or *SL problem* for short.

Because L is self-adjoint under these boundary conditions, we know from Corollary 2.19 that, if they exist, the eigenvalues of Equation (2.34) are real and its eigenfunctions are orthogonal in $\mathcal{L}^2_\rho(a, b)$. When the interval (a, b) is bounded and p does not vanish on $[a, b]$, the system of equations (2.34) to (2.36) is called a regular SL problem, otherwise it is singular. In the regular problem there is no loss of generality in assuming that $p(x) > 0$ on $[a, b]$. The solutions of the SL problem are clearly the eigenfunctions of the operator $-L/\rho$ in C^2 which satisfy the boundary conditions (2.35) and (2.36).

Now we show that the regular SL problem not only has solutions, but that there are enough of them to span $\mathcal{L}^2_\rho(a, b)$. This fundamental result is proved in stages. Assuming, for the sake of simplicity, that $\rho(x) = 1$, we first construct Green's function for the operator L under the boundary conditions (2.35) and (2.36). This choice of ρ avoids some distracting details without obscuring the general principle. We then use Green's function to arrive at an integral expression for L^{-1}, the inverse of L. The eigenfunctions of $-L$,

$$Lu + \lambda u = 0,$$

coincide with those of $-L^{-1}$,

$$L^{-1}u + \mu u = 0,$$

and the corresponding eigenvalues are related by $\mu = 1/\lambda$. We arrive at the desired conclusions by analysing the spectral properties of L^{-1}, that is, its eigenvalues and eigenfunctions.

We shall have more to say about the singular problem at the end of this section.

2.4.1 Existence of Eigenfunctions

Green's function for the self-adjoint differential operator

$$L = p\frac{d^2}{dx^2} + p'\frac{d}{dx} + r,$$

under the boundary conditions

$$\begin{aligned} \alpha_1 u(a) + \alpha_2 u'(a) &= 0, \qquad |\alpha_1| + |\alpha_2| > 0, \\ \beta_1 u(b) + \beta_2 u'(b) &= 0, \qquad |\beta_1| + |\beta_2| > 0, \end{aligned}$$

is a function

$$G : [a, b] \times [a, b] \to \mathbb{R}$$

with the following properties.

1. G is symmetric, in the sense that

$$G(x, \xi) = G(\xi, x) \qquad \text{for all } x, \xi \in [a, b],$$

and G satisfies the boundary conditions in each variable x and ξ.

2. G is a continuous function on the square $[a, b] \times [a, b]$ and of class C^2 on $[a, b] \times [a, b]\backslash\{(x, \xi) : x = \xi\}$, where it satisfies the differential equation

$$L_x G(x, \xi) = p(x)G_{xx}(x, \xi) + p'(x)G_x(x, \xi) + r(x)G(x, \xi) = 0.$$

3. The derivative $\partial G/\partial x$ has a jump discontinuity at $x = \xi$ (see Figure 2.3) given by

$$\begin{aligned}\frac{\partial G}{\partial x}(\xi^+,\xi) - \frac{\partial G}{\partial x}(\xi^-,\xi) &= \lim_{\delta\to 0^+}\left[\frac{\partial G}{\partial x}(\xi+\delta,\xi) - \frac{\partial G}{\partial x}(\xi-\delta,\xi)\right] \\ &= \frac{1}{p(\xi)}. \end{aligned} \tag{2.37}$$

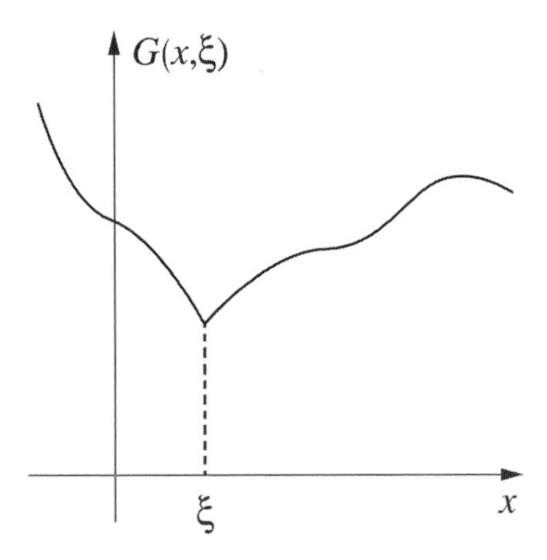

Figure 2.3 Green's function.

We assume that the homogeneous equation $Lu = 0$, under the separated boundary conditions (2.35) and (2.36), has only the trivial solution $u = 0$. This amounts to the assumption that 0 is not an eigenvalue of $-L$. There is no loss of generality in this assumption, for if $\tilde{\lambda}$ is a real number which is not an eigenvalue of $-L$ then we can define the operator

$$\tilde{L} = L + \tilde{\lambda} = p\frac{d^2}{dx^2} + p'\frac{d}{dx} + \tilde{r},$$

where $\tilde{r}(x) = r(x) + \tilde{\lambda}$. Now, in as much as

$$Lu + \lambda u = \tilde{L}u + (\lambda - \tilde{\lambda})u,$$

we see that λ is an eigenvalue of $-L$ associated with the eigenfunction u if and only if $\lambda - \tilde{\lambda}$ is an eigenvalue of $-\tilde{L}$ associated with the same eigenfunction u. Because $\tilde{\lambda}$ is not an eigenvalue of $-L$, 0 cannot be an eigenvalue of $\tilde{L}$. That there are real numbers which are not eigenvalues of $-L$ follows from the next lemma.

Lemma 2.22

The eigenvalues of $-L$ are bounded below by a real constant.

Proof

For any $u \in C^2([a,b])$ which satisfies the boundary conditions (2.35) and (2.36), we have

$$\begin{aligned}\langle -Lu, u\rangle &= \int_a^b [-(pu')'\bar{u} - r\,|u|^2]dx \\ &= \int_a^b [p\,|u'|^2 - r\,|u|^2]dx + p(a)u'(a)u(a) - p(b)u'(b)u(b) \\ &= \int_a^b [p\,|u'|^2 - r\,|u|^2]dx + p(b)\frac{\beta_1}{\beta_2}u^2(b) - p(a)\frac{\alpha_1}{\alpha_2}u^2(a).\end{aligned}$$

If $\beta_2 = 0$ then the second boundary condition implies $u(b) = 0$ and the second term on the right-hand side drops out. Similarly the third term is 0 if $\alpha_2 = 0$. The case where u is an eigenfunction of $-L$ with boundary values $u(a) = u(b) = 0$ immediately yields

$$\begin{aligned}\lambda\,\|u\|^2 &= \int_a^b p(x)\,|u'(x)|^2\,dx - \int_a^b r(x)\,|u(x)|^2\,dx \\ &\geq -\,\|u\|^2 \max\{|r(x)| : a \leq x \leq b\},\end{aligned} \tag{2.38}$$

hence $\ell = -\max\{|r(x)| : a \leq x \leq b\}$ is a lower bound of λ.

On the other hand, if u satisfies the separated boundary conditions $\alpha_1 u(a) + \alpha_2 u'(a) = 0$, $\beta_1 u(b) + \beta_2 u'(b) = 0$, then the following dimensional argument shows that there can be no more than two linearly independent eigenfunctions of $-L$ with eigenvalues less than ℓ.

Seeking a contradiction, suppose $-L$ has three linearly independent eigenfunctions u_1, u_2, and u_3 with their corresponding eigenvalues λ_1, λ_2, and λ_3 all less than ℓ. We can assume, without loss of generality, that the eigenfunctions are orthonormal. Since

$$\begin{aligned}\alpha_1 u_i(a) + \alpha_2 u_i'(a) &= 0, \\ \beta_1 u_i(b) + \beta_2 u_i'(b) &= 0, \qquad i = 1,2,3,\end{aligned}$$

we see that each of the six vectors $(u_i(a), u_i'(a))$ and $(u_i(b), u_i'(b))$ lies in a one-dimensional subspace of $\mathbb{R}^2$. Therefore the three vectors $\mathbf{u}_i = (u_i(a), u_i'(a), u_i(b), u_i'(b))$ lie in a two-dimensional subspace of $\mathbb{R}^4$. We can therefore form a linear combination $c_1\mathbf{u}_1 + c_2\mathbf{u}_2 + c_3\mathbf{u}_3$, where not all the coefficients are zeros, such that

$$c_1\mathbf{u}_1 + c_2\mathbf{u}_2 + c_3\mathbf{u}_3 = \mathbf{0}.$$

But this implies

$$c_1u_1(a) + c_2u_2(a) + c_3u_3(a) = 0,$$
$$c_1u_1(b) + c_2u_2(b) + c_3u_3(b) = 0.$$

The function

$$v(x) = c_1u_1(x) + c_2u_2(x) + c_3u_3(x)$$

is therefore an eigenfunction of $-L$ which satisfies $v(a) = v(b) = 0$, and consequently its eigenvalue is bounded below by ℓ. But this is contradicted by the inequality

$$\langle -Lv, v\rangle = \lambda_1 |c_1|^2 + \lambda_2 |c_2|^2 + \lambda_3 |c_3|^2 < \ell(|c_1|^2 + |c_2|^2 + |c_3|^2) = \ell \|v\|^2 .$$

□

Now we construct Green's function for the operator L under the boundary conditions (2.35) and (2.36),

$$\alpha_1u(a) + \alpha_2u'(a) = 0, \qquad |\alpha_1| + |\alpha_2| > 0,$$
$$\beta_1u(b) + \beta_2u'(b) = 0, \qquad |\beta_1| + |\beta_2| > 0.$$

According to a standard existence theorem for second-order differential equations [6], $Lu = 0$ has two unique (nontrivial) solutions v_1 and v_2 such that

$$v_1(a) = \alpha_2, \qquad v_1'(a) = -\alpha_1,$$
$$v_2(b) = \beta_2, \qquad v_2'(b) = -\beta_1.$$

Thus v_1 satisfies the first boundary condition at $x = a$,

$$\alpha_1v_1(a) + \alpha_2v_1'(a) = 0,$$

and v_2 satisfies the second condition

$$\beta_1v_2(b) + \beta_2v_2'(b) = 0.$$

Clearly v_1 and v_2 are linearly independent, otherwise each would be a (nonzero) constant multiple of the other, and both would then satisfy $Lu = 0$ and the boundary conditions (2.35) and (2.36). But this would contradict the assumption that 0 is not an eigenvalue of L.

Now we define

$$G(x,\xi) = \begin{cases} c^{-1}v_1(\xi)v_2(x), & a \le \xi \le x \le b \\ c^{-1}v_1(x)v_2(\xi), & a \le x \le \xi \le b, \end{cases} \tag{2.39}$$

where

$$c = p(x)[v_1(x)v_2'(x) - v_1'(x)v_2(x)] = p(x)W(v_1, v_2)(x) \tag{2.40}$$

is a nonzero constant. This follows from the fact that neither p nor W vanishes on $[a,b]$, and the Lagrange identity (see Exercise 2.16)

$$[p(v_1v_2' - v_1'v_2)]' = v_1Lv_2 - v_2Lv_1 = 0 \text{ on } [a,b].$$

It is now a simple matter to show that all the properties of $G(\xi, x)$ listed above are satisfied. Here we verify Equation (2.37). Differentiating (2.39) we obtain

$$\frac{\partial G}{\partial \xi}(x, x+\delta) - \frac{\partial G}{\partial \xi}(x, x-\delta) = \frac{1}{c}[v_1(x)v_2'(x+\delta) - v_1'(x-\delta)v_2(x)],$$

where $\delta > 0$. In view of (2.40) and the continuity of v_1' and v_2', this expression tends to $1/p(x)$ as $\delta \to 0$.

Now we define the operator T on $C([a,b])$ by

$$(Tf)(x) = \int_a^b G(x,\xi)f(\xi)d\xi, \tag{2.41}$$

and show that the function Tf is of class $C^2([a,b])$ and solves the differential equation $Lu = f$. Rewriting (2.41) and differentiating,

$$(Tf)(x) = \int_a^x G(x,\xi)f(\xi)d\xi + \int_x^b G(x,\xi)f(\xi)d\xi,$$

$$(Tf)'(x) = \int_a^x G_x(x,\xi)f(\xi)d\xi + \int_x^b G_x(x,\xi)f(\xi)d\xi,$$

$$(Tf)''(x) = \int_a^x G_{xx}(x,\xi)f(\xi)d\xi + G_x(x,x^-)f(x^-)$$

$$+ \int_x^b G_{xx}(x,\xi)f(\xi)d\xi - G_x(x,x^+)f(x^+),$$

where we have used the continuity of G and f at $\xi = x$ to obtain $(Tf)'(x)$. Because

$$G_x(x,x^-) - G_x(x,x^+) = \frac{1}{c}[v_1(x^-)v_2'(x) - v_1'(x)v_2(x^+)] = \frac{1}{p(x)},$$

by the continuity of v_1 and v_2, we also have

$$(Tf)''(x) = \int_a^x G_{xx}(x,\xi)f(\xi)d\xi + \int_x^b G_{xx}(x,\xi)f(\xi)d\xi + \frac{f(x)}{p(x)},$$

from which we conclude that Tf lies in $C^2([a,b])$ and satisfies

$$\begin{aligned}L(Tf)(x) &= p(x)(Tf)''(x) + p'(x)(Tf)'(x) + r(x)(Tf)(x)\\ &= \int_a^x L_x G(x,\xi) f(\xi) d\xi + \int_x^b L_x G(x,\xi) f(\xi) d\xi + f(x)\\ &= f(x),\end{aligned}$$

in view of the fact that $L_x G(x,\xi) = 0$ for all $\xi \neq x$.

From property (1) of G it is clear that Tf also satisfies the boundary conditions (2.35) and (2.36). On the other hand, if $u \in C^2([a,b])$ satisfies the boundary conditions (2.35) and (2.36), then we can integrate by parts and use the continuity of p, u, u' and the properties of G to obtain

$$\begin{aligned}T(Lu)(x) &= \int_a^x G(x,\xi) Lu(\xi) d\xi + \int_x^b G(x,\xi) Lu(\xi) d\xi\\ &= p(\xi)\left[u'(\xi)G(x,\xi) - u(\xi)G_\xi(x,\xi)\right]\big|_a^x + \int_a^x u(\xi) L_\xi G(x,\xi) d\xi\\ &\quad + p(\xi)\left[u'(\xi)G(x,\xi) - u(\xi)G_\xi(x,\xi)\right]\big|_x^b + \int_x^b u(\xi) L_\xi G(x,\xi) d\xi\\ &= p(\xi)u(\xi)G_\xi(x,\xi)\big|_{x^-}^{x^+} + p(\xi)\left[u'(\xi)G(x,\xi) - u(\xi)G_\xi(x,\xi)\right]\big|_a^b\\ &= u(x),\end{aligned}$$

where we have used Equation (2.37) and the fact that u and G satisfy the separated boundary conditions. The operator T therefore acts as a sort of inverse to the operator L, and the SL system of equations

$$\begin{aligned}Lu + \lambda u &= 0,\\ \alpha_1 u(a) + \alpha_2 u'(a) &= 0,\\ \beta_1 u(b) + \beta_2 u'(b) &= 0,\end{aligned}$$

is equivalent to the single eigenvalue equation

$$Tu = \mu u,$$

where $\mu = -1/\lambda$. In other words, u is an eigenfunction of the SL problem associated with the eigenvalue λ if and only if it is an eigenfunction of T associated with the eigenvalue $-1/\lambda$. We shall deduce the spectral properties of the regular SL problem from those of the integral operator T.

The following example illustrates the method we have described for constructing Green's function in the special case where the operator L is d^2/dx^2 on the interval $[0,1]$.

Example 2.23

Let

$$u'' + \lambda u = 0, \qquad 0 < x < 1,$$
$$u(0) = u(1) = 0.$$

We have already seen in Example 2.16 that this system of equations has only positive eigenvalues. The general solution of the differential equation is

$$v(x) = c_1 \cos\sqrt{\lambda}x + c_2 \sin\sqrt{\lambda}x.$$

Noting that $\alpha_1 = \beta_1 = 1$ and $\alpha_2 = \beta_2 = 0$, we seek the solution v_1 which satisfies the following initial conditions at $x = 0$,

$$v_1(0) = \alpha_2 = 0, \qquad v_1'(0) = -\alpha_1 = -1.$$

These conditions imply

$$c_1 = 0, \qquad \sqrt{\lambda}c_2 = -1.$$

Hence

$$v_1(x) = -\frac{1}{\sqrt{\lambda}} \sin\sqrt{\lambda}x.$$

Similarly, the solution v_2 which satisfies the conditions

$$v_2(1) = \beta_2 = 0, \qquad v_2'(1) = -\beta_1 = -1,$$

is readily found to be

$$v_2(x) = \frac{\sin\sqrt{\lambda}}{\sqrt{\lambda}} \cos\sqrt{\lambda}x - \frac{\cos\sqrt{\lambda}}{\sqrt{\lambda}} \sin\sqrt{\lambda}x.$$

Now Green's function, according to (2.39), is given by

$$G(x,\xi) = \begin{cases} c^{-1}v_1(\xi)v_2(x), & 0 \le \xi \le x \le 1, \\ c^{-1}v_1(x)v_2(\xi), & 0 \le x \le \xi \le 1. \end{cases}$$

The constant c can be computed using (2.39), and is given by $\sin\sqrt{\lambda}/\sqrt{\lambda}$. Thus

$$G(x,\xi) = \begin{cases} \dfrac{\sin\sqrt{\lambda}\xi}{\sqrt{\lambda}\sin\sqrt{\lambda}} \left(\cos\sqrt{\lambda}\sin\sqrt{\lambda}x - \sin\sqrt{\lambda}\cos\sqrt{\lambda}x\right), & 0 \le \xi \le x \le 1, \\ \dfrac{\sin\sqrt{\lambda}x}{\sqrt{\lambda}\sin\sqrt{\lambda}} \left(\cos\sqrt{\lambda}\sin\sqrt{\lambda}\xi - \sin\sqrt{\lambda}\cos\sqrt{\lambda}\xi\right), & 0 \le x \le \xi \le 1. \end{cases}$$

Note that

$$\begin{aligned} G_\xi(x,x^+) - G_\xi(x,x^-) &= \frac{\sin\sqrt{\lambda}x}{\sin\sqrt{\lambda}}\left(\cos\sqrt{\lambda}\cos\sqrt{\lambda}x^+ + \sin\sqrt{\lambda}\sin\sqrt{\lambda}x^+\right) \\ &\quad - \frac{\cos\sqrt{\lambda}x^-}{\sin\sqrt{\lambda}}\left(\cos\sqrt{\lambda}\sin\sqrt{\lambda}x - \sin\sqrt{\lambda}\cos\sqrt{\lambda}x\right) \\ &= 1. \end{aligned}$$

The other properties of G can easily be checked.

An infinite set F of functions defined and continuous on $[a,b]$ is said to be *equicontinuous* on $[a,b]$ if, given any $\varepsilon > 0$, there is a $\delta > 0$, which depends only on ε, such that

$$x,\xi \in [a,b], \qquad |x-\xi| < \delta \;\Rightarrow\; |f(x) - f(\xi)| < \varepsilon \qquad \text{for all } f \in F.$$

This condition clearly implies that every member of an equicontinuous set of functions is a uniformly continuous function on $[a,b]$; but, more than that, the same δ works for all the functions in F. The set F is *uniformly bounded* if there is a positive number M such that

$$|f(x)| \le M \quad \text{for all } f \in F,\ x \in [a,b].$$

According to the Ascoli–Arzela theorem [14], if F is an infinite, uniformly bounded, and equicontinuous set of functions on the bounded interval $[a,b]$, then F contains a sequence $(f_n : n \in \mathbb{N})$ which is uniformly convergent on $[a,b]$. The limit of f_n, by Theorem 1.17, is necessarily a continuous function on $[a,b]$.

Lemma 2.24

Let T be the integral operator defined by Equation (2.41). The set of functions $\{Tu\}$, where $u \in C([a,b])$ and $\|u\| \le 1$, is uniformly bounded and equicontinuous.

Proof

Green's function G is continuous on the closed square $[a,b] \times [a,b]$, therefore $|G(x,\xi)|$ is uniformly continuous and bounded by some positive constant M. From (2.41) and the CBS inequality,

$$|Tu(x)| = |\langle G(x,\xi), u(\xi)\rangle| \le M\sqrt{b-a}\,\|u\|,$$

hence the set $\{|Tu| : \|u\| \leq 1\}$ is bounded above by $M\sqrt{b-a}$. Because G is uniformly continuous on the square $[a,b] \times [a,b]$, given any $\varepsilon > 0$, there is a $\delta > 0$ such that

$$x_1, x_2 \in [a,b], \ |x_2 - x_1| < \delta \Rightarrow |G(x_2,\xi) - G(x_1,\xi)| < \varepsilon \quad \text{for all } \xi \in [a,b].$$

If u is continuous on $[a,b]$, then

$$|x_2 - x_1| < \delta \ \Rightarrow \ |Tu(x_2) - Tu(x_1)| \leq \varepsilon\sqrt{b-a}\,\|u\|\,.$$

Thus the functions $\{Tu : u \in C[a,b], \|u\| \leq 1\}$ are equicontinuous. □

The norm of the operator T, denoted by $\|T\|$, is defined by

$$\|T\| = \sup\{\|Tu\| : u \in C([a,b]), \|u\| = 1\}. \tag{2.42}$$

From this it follows that, if $u \neq 0$, then

$$\|Tu\| = \|T(u/\|u\|)\|\,\|u\| \leq \|T\|\,\|u\|\,.$$

But because this inequality also holds when $u = 0$, it holds for all $u \in C([a,b])$. It is also a simple exercise (see Exercise 2.33) to show that

$$\|T\| = \sup_{\|u\|=1} |\langle Tu, u\rangle|\,. \tag{2.43}$$

Now we can state our first existence theorem for the eigenvalue problem under consideration.

Theorem 2.25

Either $\|T\|$ or $-\|T\|$ is an eigenvalue of T.

Proof

$\|T\| = \sup\{|\langle Tu, u\rangle| : u \in C([a,b]), \|u\| = 1\}$ and $\langle Tu, u\rangle$ is a real number, hence either $\|T\| = \sup\langle Tu, u\rangle$ or $\|T\| = -\inf\langle Tu, u\rangle\,,\ \|u\| = 1$. Suppose $\|T\| = \sup\langle Tu, u\rangle\,.$ Then there is a sequence of functions $u_k \in C([a,b])$, with $\|u_k\| = 1$, such that

$$\langle Tu_k, u_k\rangle \to \|T\| \quad \text{as } k \to \infty.$$

Since $\{Tu_k\}$ is uniformly bounded and equicontinuous, we can use the Ascoli–Arzela theorem to conclude that it has a subsequence $\{Tu_{k_j}\}$ which is uniformly convergent on $[a,b]$ to a continuous function φ_0. We now prove that φ_0 is an eigenfunction of T corresponding to the eigenvalue $\mu_0 = \|T\|\,.$

As $j \to \infty$ we have

$$\sup_{x\in[a,b]} \left|Tu_{k_j}(x) - \varphi_0(x)\right| \to 0,$$

which implies

$$\left\|Tu_{k_j} - \varphi_0\right\| \to 0, \tag{2.44}$$

and therefore $\left\|Tu_{k_j}\right\| \to \|\varphi_0\|$. Furthermore, because $\left\langle Tu_{k_j}, u_{k_j}\right\rangle \to \mu_0$,

$$\left\|Tu_{k_j} - \mu_0 u_{k_j}\right\|^2 = \left\|Tu_{k_j}\right\|^2 + \mu_0^2 - 2\mu_0 \left\langle Tu_{k_j}, u_{k_j}\right\rangle \to \|\varphi_0\|^2 - \mu_0^2, \tag{2.45}$$

hence $\|\varphi_0\|^2 \geq \mu_0^2 > 0$, and the function φ_0 cannot vanish identically on $[a, b]$. Because $\left\|Tu_{k_j}\right\|^2 \leq \|T\|^2 \left\|u_{k_j}\right\|^2 = \mu_0^2$, it also follows from (2.45) that

$$0 \leq \left\|Tu_{k_j} - \mu_0 u_{k_j}\right\|^2 \leq 2\mu_0^2 - 2\mu_0 \left\langle Tu_{k_j}, u_{k_j}\right\rangle.$$

But $\left\langle Tu_{k_j}, u_{k_j}\right\rangle \to \mu_0$, hence

$$\left\|Tu_{k_j} - \mu_0 u_{k_j}\right\| \to 0. \tag{2.46}$$

Now we can write

$$\begin{aligned} 0 &\leq \|T\varphi_0 - \mu_0\varphi_0\| \\ &\leq \left\|T\varphi_0 - T(Tu_{k_j})\right\| + \left\|T(Tu_{k_j}) - \mu_0 Tu_{k_j}\right\| + \left\|\mu_0 Tu_{k_j} - \mu_0\varphi_0\right\|. \end{aligned}$$

In the limit as $j \to \infty$, using the inequality $\|Tu\| \leq \|T\| \|u\|$ together with (2.44) and (2.46), we conclude that $\|T\varphi_0 - \mu_0\varphi_0\| = 0$. $T\varphi_0 - \mu_0\varphi_0$ being continuous, this implies $T\varphi_0(x) = \mu_0\varphi_0(x)$ for all $x \in [a, b]$.

If $\|T\| = -\inf \langle Tu, u\rangle$, a similar argument leads to the same conclusion. □

Let

$$\begin{aligned} \psi_0 &= \frac{\varphi_0}{\|\varphi_0\|}, \\ G_1(x, \xi) &= G(x, \xi) - \mu_0\psi_0(x)\bar{\psi}_0(\xi), \\ (T_1 u)(x) &= \int_a^b G_1(x, \xi)u(\xi)d\xi \\ &= Tu(x) - \mu_0 \langle u, \psi_0\rangle \psi_0(x) \qquad \text{for all } u \in C([a, b]). \end{aligned} \tag{2.47}$$

The function G_1 has the same regularity and symmetry properties as G, hence Lemma 2.24 and Theorem 2.25 clearly apply to the self-adjoint operator T_1. If $\|T_1\| \neq 0$ then we define

$$\sup\{|\langle T_1 u, u\rangle| : u \in C([a, b]), \|u\| = 1\} = |\mu_1|,$$

where μ_1 is a (nonzero) real number which, by Theorem 2.25, is an eigenvalue of T_1 corresponding to some (nonzero) eigenfunction $\varphi_1 \in C([a,b])$; that is,

$$T_1\varphi_1 = \mu_1\varphi_1.$$

Let $\psi_1 = \varphi_1/\|\varphi_1\|$ Then, for any $u \in C([a,b])$,

$$\begin{aligned}\langle T_1 u, \psi_0\rangle &= \langle Tu, \psi_0\rangle - \mu_0 \langle\langle u, \psi_0\rangle \psi_0, \psi_0\rangle \\ &= \langle u, T\psi_0\rangle - \mu_0 \langle u, \psi_0\rangle \\ &= 0\end{aligned}$$

since $T\psi_0 = \mu_0\psi_0$. In particular, $\langle T_1\psi_1, \psi_0\rangle = \langle \mu_1\psi_1, \psi_0\rangle = 0$, and therefore ψ_1 is orthogonal to ψ_0. Now Equation (2.47) gives

$$T\psi_1 = T_1\psi_1 = \mu_1\psi_1.$$

Thus ψ_1 is also an eigenfunction of T, and the associated eigenvalue satisfies

$$|\mu_1| = \|T\psi_1\| \leq \|T\| = |\mu_0|.$$

Setting

$$G_2(x,\xi) = G_1(x,\xi) - \mu_1\psi_1(x)\bar{\psi}_1(\xi) = G(x,\xi) - \sum_{k=0}^{1} \mu_k\psi_k(x)\bar{\psi}_k(\xi)$$

$$T_2 u = T_1 u - \mu_1 \langle u, \psi_1\rangle \psi_1 = Tu - \sum_{k=0}^{1} \mu_k \langle u, \psi_k\rangle \psi_k,$$

and proceeding as above, we deduce the existence of a third normalized eigenfunction ψ_2 of T associated with a real eigenvalue μ_2 such that ψ_2 is orthogonal to both ψ_0 and ψ_1 and $|\mu_2| \leq |\mu_1|$. Thus we obtain an orthonormal sequence of eigenfunctions $\psi_0, \psi_1, \psi_2, \ldots$ of T corresponding to the sequence of eigenvalues $|\mu_0| \geq |\mu_1| \geq |\mu_2| \geq \ldots$. The sequence of eigenfunctions terminates only if $\|T_n\| = 0$ for some n. In that case

$$0 = LT_n u = LTu - \sum_{k=0}^{n-1} \mu_k \langle u, \psi_k\rangle L\psi_k = u - \sum_{k=0}^{n-1} \mu_k \langle u, \psi_k\rangle L\psi_k$$

for all $u \in C([a,b])$, which would yield

$$u = \sum_{k=0}^{n-1} \mu_k \langle u, \psi_k\rangle L\psi_k = \sum_{k=0}^{n-1} \langle u, \psi_k\rangle LT\psi_k = \sum_{k=0}^{n-1} \langle u, \psi_k\rangle \psi_k.$$

But because no finite set of eigenfunctions can span $C([a,b])$, this last equality cannot hold for all $u \in C([a,b])$. Consequently, $\|T_n\| > 0$ for all $n \in \mathbb{N}$, and we have proved the following.

Theorem 2.26 (Existence Theorem)

The integral operator T defined by Equation (2.41) has an infinite sequence of eigenfunctions (ψ_n) which are orthonormal in $\mathcal{L}^2(a,b)$.

2.4.2 Completeness of the Eigenfunctions

If f is any function in $\mathcal{L}^2(a,b)$ then, by Bessel's inequality (1.22), we have

$$\sum_{k=0}^{\infty} |\langle f, \psi_k \rangle|^2 \leq \|f\|^2 .$$

To prove the completeness of the eigenfunctions (ψ_n) in $\mathcal{L}^2(a,b)$ we have to show that this inequality is in fact an equality. This we do by first proving that

$$f = \sum_{k=0}^{\infty} \langle f, \psi_k \rangle \psi_k$$

for any $f \in C^2([a,b])$ which satisfies the boundary conditions (2.35) and (2.36), and then using the density of $C^2([a,b])$ in $\mathcal{L}^2(a,b)$ to extend this equality to $\mathcal{L}^2(a,b)$.

Theorem 2.27

Given any function $f \in C^2([a,b])$ which satisfies the boundary conditions (2.35) and (2.36), the infinite series $\sum \langle f, \psi_k \rangle \psi_k$ converges uniformly to f on $[a,b]$.

Proof

For every fixed $x \in [a,b]$, we have

$$\langle G(x,\cdot), \psi_k \rangle = T\bar{\psi}_k(x) = \mu_k \bar{\psi}_k(x) \quad \text{for all } k \in \mathbb{N}.$$

Bessel's inequality, applied to G as a function of ξ, yields

$$\sum_{k=0}^{n} \mu_k^2 |\psi_k(x)|^2 \leq \int_a^b |G(x,\xi)|^2 \, d\xi \quad \text{for all } x \in [a,b], \qquad n \in \mathbb{N}.$$

Integrating with respect to x and letting $n \to \infty$, we obtain

$$\sum_{k=0}^{\infty} \mu_k^2 \leq M^2 (b-a)^2, \tag{2.48}$$

where M is the maximum value of $|G(x,\xi)|$ on the square $[a,b]\times[a,b]$. An immediate consequence of the inequality (2.48) is that

$$\lim_{n\to\infty}|\mu_n| = 0. \tag{2.49}$$

With

$$G_n(x,\xi) = G(x,\xi) - \sum_{k=0}^{n-1}\mu_k\psi_k(x)\bar{\psi}_k(\xi),$$

we have already seen that the integral operator

$$T_n : u(x) \mapsto \int_a^b G_n(x,\xi)u(\xi)d\xi, \qquad u\in C([a,b]),$$

has eigenvalue μ_n and norm

$$\|T_n\| = |\mu_n|\,.$$

Therefore, for any $u \in C([a,b])$,

$$\|T_n u\| = \left\| Tu - \sum_{k=0}^{n-1}\mu_k\langle u,\psi_k\rangle\psi_k\right\| \leq |\mu_n|\,\|u\| \to 0 \quad \text{as } n\to\infty \tag{2.50}$$

in view of (2.49). If $n > m$, then

$$\sum_{k=m}^{n}\mu_k\langle u,\psi_k\rangle\psi_k = T\left(\sum_{k=m}^{n}\langle u,\psi_k\rangle\psi_k\right);$$

but because $|Tu| \leq M\sqrt{b-a}\,\|u\|$ for all $u\in C([a,b])$, it follows that

$$\begin{aligned}\left|\sum_{k=m}^{n}\mu_k\langle u,\psi_k\rangle\psi_k\right| &\leq \|T\|\left\|\sum_{k=m}^{n}\langle u,\psi_k\rangle\psi_k\right\| \\ &\leq M\sqrt{b-a}\left(\sum_{k=m}^{n}|\langle u,\psi_k\rangle|^2\right)^{1/2}.\end{aligned}$$

By Bessel's inequality, the right-hand side of this inequality tends to zero as $m,\ n\to\infty$, hence the series

$$\sum_{k=0}^{\infty}\mu_k\langle u,\psi_k\rangle\psi_k$$

converges uniformly on $[a,b]$ to a continuous function. The continuity of Tu and (2.50) now imply

$$Tu(x) = \sum_{k=0}^{\infty}\mu_k\langle u,\psi_k\rangle\psi_k(x) \quad \text{for all } x\in[a,b]. \tag{2.51}$$

If $f \in C^2([a,b])$ satisfies the boundary conditions (2.35) and (2.36), then $u = Lf$ is a continuous function on $[a,b]$ and $f = Tu$. Since

$$\mu_k \langle u, \psi_k \rangle = \langle u, \mu_k \psi_k \rangle = \langle u, T\psi_k \rangle = \langle Tu, \psi_k \rangle = \langle f, \psi_k \rangle,$$

Equation (2.51) yields

$$f(x) = \sum_{k=0}^{\infty} \langle f, \psi_k \rangle \psi_k(x) \qquad \text{for all } x \in [a,b].$$

□

Now we need to prove that any function f in $\mathcal{L}^2(a,b)$ can be approximated (in the $\mathcal{L}^2$ norm) by a C^2 function on $[a,b]$ which satisfies the separated boundary conditions (2.35) and (2.36). In fact it suffices to prove the density of $C^2([a,b])$ in $\mathcal{L}^2(a,b)$ regardless of the boundary conditions; for if $g \in C^2([a,b])$ then we can form a uniformly bounded sequence of functions $g_n \in C^2([a,b])$ such that $g_n(x) = g(x)$ on $[a+1/n, b-1/n]$ and $g_n(a) = g_n'(a) = g_n(b) = g_n'(b) = 0$. Such a sequence clearly satisfies the boundary conditions and yields

$$\lim_{n\to\infty} \|g - g_n\| = 0.$$

If $f \in \mathcal{L}^2(a,b)$, we know from Theorem 1.27 that, given any $\varepsilon > 0$, there is a function h, continuous on $[a,b]$, such that

$$\|f - h\| < \varepsilon. \tag{2.52}$$

Hence we have to prove that, for any $h \in C([a,b])$, there is a function $g \in C^2([a,b])$ such that

$$\|h - g\| < \varepsilon. \tag{2.53}$$

This can be deduced from the Weierstrass approximation theorem (see [1] or [14]), but here is a direct proof of this result.

Define the positive C^∞ function

$$\alpha(x) = \begin{cases} c \exp\left(\dfrac{1}{x^2 - 1}\right), & |x| < 1 \\ 0, & |x| \geq 1, \end{cases}$$

where

$$c = \left[\int_{-\infty}^{\infty} \exp\left(\frac{1}{x^2-1}\right) dx\right]^{-1},$$

so that $\int_{-\infty}^{\infty} \alpha(x)dx = 1$. Then we define the sequence of functions

$$\alpha_n(x) = n\alpha(nx), \qquad n \in \mathbb{N},$$

which also lie in $C^\infty(\mathbb{R})$. Because α vanishes on $|x| \geq 1$, each function α_n vanishes on $|x| \geq 1/n$, and satisfies

$$\int_{-\infty}^{\infty} \alpha_n(x)dx = \int_{-\infty}^{\infty} \alpha(x)dx = 1 \quad \text{for all } n.$$

Finally, we extend h as a continuous function from $[a, b]$ to $\mathbb{R}$ by setting

$$h(x) = \begin{cases} 0, & x < a-1,\ x > b+1 \\ h(a)(x-a+1), & a-1 \leq x < a \\ h(b)(-x+b+1), & b < x \leq b+1, \end{cases}$$

and define the convolution of h and α_n by

$$\begin{aligned} (h * \alpha_n)(x) &= \int_{-\infty}^{\infty} h(y)\alpha_n(x-y)dy \\ &= \int_{x-(1/n)}^{x+(1/n)} h(y)\alpha_n(x-y)dy \\ &= \int_{-1/n}^{1/n} h(x-y)\alpha_n(y)dy. \end{aligned}$$

With $\alpha_n \in C^\infty$ the function $h * \alpha_n$ is also a C^∞ function on $\mathbb{R}$. Furthermore,

$$\begin{aligned} |h(x) - (h * \alpha_n)(x)| &= \left| \int_{-1/n}^{1/n} [h(x-y)\alpha_n(y)dy - h(x)\alpha_n(y)]dy \right| \\ &\leq \sup\{|h(x-y) - h(x)| : x \in [a,b], |y| \leq 1/n\}. \end{aligned}$$

Because h is continuous on $\mathbb{R}$, it is uniformly continuous on any finite real interval, and therefore, as $n \to \infty$,

$$|h(x) - (h * \alpha_n)(x)| \leq \sup\{|h(x-y) - h(x)| : x \in [a,b], |y| \leq 1/n\} \to 0.$$

We have actually shown here that C^∞, which is a subset of C^2, is dense in $C([a,b])$ in the supremum norm, which is more than we set out to do. Consequently, there is an integer N such that

$$\|h - h * \alpha_n\| < \varepsilon \quad \text{for all } n \geq N.$$

By choosing the function g in (2.53) to be $h * \alpha_N$, we obtain

$$\|f - g\| \leq \|f - h\| + \|h - g\| < 2\varepsilon \tag{2.54}$$

for any $f \in \mathcal{L}^2(a,b)$. Using the triangle and Bessel inequalities, we also have

$$\begin{aligned} \left\| f - \sum_{k=0}^{n} \langle f, \psi_k \rangle \psi_k \right\| &\leq \|f - g\| + \left\| g - \sum_{k=0}^{n} \langle g, \psi_k \rangle \psi_k \right\| + \left\| \sum_{k=0}^{n} \langle g - f, \psi_k \rangle \psi_k \right\| \\ &\leq 2\|f - g\| + \left\| g - \sum_{k=0}^{n} \langle g, \psi_k \rangle \psi_k \right\|. \end{aligned} \tag{2.55}$$

In view of Theorem 2.27, we can make $\|g - \sum_{k=0}^{n} \langle g, \psi_k \rangle \psi_k\| < \varepsilon$ by choosing n large enough, say $n \geq M$. This, together with (2.54) and (2.55), now imply

$$\left\| f - \sum_{k=0}^{n} \langle f, \psi_k \rangle \psi_k \right\| < 5\varepsilon \qquad \text{for all } n \geq \max\{N, M\}.$$

We have therefore proved the following.

Theorem 2.28 (Completeness Theorem)

If $f \in \mathcal{L}^2(a, b)$, then

$$\lim_{n\to\infty} \left\| f - \sum_{k=0}^{n} \langle f, \psi_k \rangle \psi_k \right\| = 0. \tag{2.56}$$

Equation (2.56) is equivalent to the identity

$$f = \sum_{k=0}^{\infty} \langle f, \psi_k \rangle \psi_k,$$

where the equality is in $\mathcal{L}^2(a, b)$. It is also equivalent to Parseval's relation

$$\|f\|^2 = \sum_{k=0}^{\infty} |\langle f, \psi_k \rangle|^2 .$$

Each one these equations expresses the fact that the sequence (ψ_k) of orthonormal eigenfunctions of T forms a complete set in $\mathcal{L}^2(a, b)$.

Going back to the SL problem

$$\begin{aligned} Lu + \lambda u &= 0, \\ \alpha_1 u(a) + \alpha_2 u'(a) &= 0, \\ \beta_1 u(b) + \beta_2 u'(b) &= 0, \end{aligned}$$

where $u \in C^2([a, b])$, we have shown that this system of equations is equivalent to the single integral equation

$$Tu = \mu u = -u/\lambda.$$

The boundary conditions determine a unique solution to the differential equation $Lu + \lambda u = 0$, therefore each eigenvalue λ of $-L$ corresponds to a unique eigenfunction u. This is equivalent to saying that each eigenvalue μ of T corresponds to a unique eigenfunction u.

By (2.49), $1/|\lambda_n| = |\mu_n| \to 0$, and it follows from Lemma 2.22 that $\lambda_n \to \infty$ as $n \to \infty$. Reintroducing the weight function ρ, we therefore arrive at the following fundamental theorem.

Theorem 2.29

Assuming $p', r, \rho \in C([a,b])$, and $p, \rho > 0$ on $[a,b]$, the SL eigenvalue problem defined by Equations (2.34) to (2.36) has an infinite sequence of real eigenvalues

$$\lambda_0 < \lambda_1 < \lambda_2 < \cdots$$

such that $\lambda_n \to \infty$. To each eigenvalue λ_n corresponds a single eigenfunction φ_n, and the sequence of eigenfunctions $(\varphi_n : n \in \mathbb{N}_0)$ forms an orthogonal basis of $\mathcal{L}^2_\rho(a,b)$.

Remark 2.30

1. If the separated boundary conditions (2.35) and (2.36) are replaced by the periodic conditions

$$u(a) = u(b), \qquad u'(a) = u'(b),$$

it is a simple matter to verify that Equation (2.26) is satisfied provided $p(a) = p(b)$ (Exercise 2.26). The operator defined by (2.33) is then self-adjoint and its eigenvalues are bounded below by $-\max\{|r(x)| : a \leq x \leq b\}$ (see Equation (2.38)). The results obtained above remain valid, except that the uniqueness of the eigenfunction for each eigenvalue is not guaranteed. This is illustrated by Example 3.1 in the next chapter. For more information on the SL problem under periodic boundary conditions see [6].

2. It can also be shown that each eigenfunction φ_n has exactly n zeros in the interval (a,b), and it is worth checking this claim against the examples which are discussed henceforth. A proof of this result may be found in [6].

3. The interested reader may wish to refer to [18] for an extensive up-to-date bibliography on the SL problem and its history since the early part of the nineteenth century.

Example 2.31

Find the eigenvalues and eigenfunctions of the equation

$$u'' + \lambda u = 0, \qquad 0 \leq x \leq l,$$

subject to one of the pairs of separated boundary conditions

$$(i)\ \ u(0) = u(l) = 0,$$
$$(ii)\ \ u'(0) = u'(l) = 0.$$

Solution

The differential operator is formally self-adjoint and the boundary conditions (i) and (ii) are homogeneous and separated, so both Theorems 2.14 and 2.29 apply. We have already seen in Example 2.16 that, under the boundary conditions (i), the equation $u'' + \lambda u = 0$ has only positive eigenvalues. Under the boundary conditions (ii) it is a simple matter to show that it has no negative eigenvalues. This is consistent with Remark 2.30 in as much as $r(x) \equiv 0$.

If $\lambda > 0$, the general solution of the differential equation is

$$u(x) = c_1 \cos \sqrt{\lambda} x + c_2 \sin \sqrt{\lambda} x. \tag{2.57}$$

Applying conditions (i), we obtain

$$\begin{aligned} u(0) &= c_1 = 0, \\ u(l) &= c_2 \sin \sqrt{\lambda} l = 0 \Rightarrow \lambda_n = \frac{n^2 \pi^2}{l^2}, \qquad n \in \mathbb{N}. \end{aligned}$$

In accordance with Theorem 2.29 the eigenvalues λ_n tend to ∞, and the corresponding eigenfunctions

$$u_n(x) = \sin \frac{n\pi}{l} x$$

are orthogonal in $\mathcal{L}^2(0, l)$, because, for all $m \neq n$,

$$\int_0^l \sin \frac{m\pi}{l} x \sin \frac{n\pi}{l} x \, dx = \frac{1}{2} \int_0^l \left[\cos(m-n) \frac{\pi}{l} x - \cos(m+n) \frac{\pi}{l} x \right] dx = 0.$$

From Theorem 2.29 we also conclude that the sequence of functions

$$\sin \frac{n\pi}{l} x, \qquad n \in \mathbb{N},$$

is complete in $\mathcal{L}^2(0, l)$.

Now we turn to the boundary conditions (ii): if $\lambda = 0$, the general solution of $u'' = 0$ is

$$u(x) = c_1 x + c_2,$$

and conditions (ii) imply $c_1 = 0$. u is therefore a constant, which we can take to be 1. Thus

$$\lambda_0 = 0, \qquad u_0(x) = 1$$

is the first eigenvalue–eigenfunction pair. But if $\lambda > 0$, the first of conditions (ii) applied to (2.57) gives

$$u'(0) = \sqrt{\lambda} c_2 = 0,$$

hence $c_2 = 0$. The condition at $x = l$ yields

$$u'(l) = -c_1\sqrt{\lambda}\sin\sqrt{\lambda}l = 0 \Rightarrow \lambda_n = \frac{n^2\pi^2}{l^2}$$
$$u_n(x) = \cos\frac{n\pi}{l}x, \qquad n \in \mathbb{N}.$$

Thus the eigenvalues and eigenfunctions under conditions (ii) are

$$\lambda_n = \frac{n^2\pi^2}{l^2}, \qquad u_n(x) = \cos\frac{n\pi}{l}x, \qquad n \in \mathbb{N}_0.$$

Again we can verify that the sequence $(\cos(n\pi/lx) : n \in \mathbb{N}_0)$ is orthogonal in $\mathcal{L}^2(0,l)$, and we conclude from Theorem 2.29 that it also spans this space.

According to Example 2.31, any function $f \in \mathcal{L}^2(0,l)$ can be represented by an infinite series of the form

$$f(x) = \sum_{n=0}^{\infty} a_n \cos\frac{n\pi}{l}x, \tag{2.58}$$

or of the form

$$f(x) = \sum_{n=1}^{\infty} b_n \sin\frac{n\pi}{l}x, \tag{2.59}$$

the equality in each case being in $\mathcal{L}^2(0,l)$. Note that these two representations of $f(x)$ do not necessarily give the same value at every point in $[0,l]$. For example, at $x = 0$, we obtain $f(0) = \sum_{n=0}^{\infty} a_n$ in the first representation and $f(0) = 0$ in the second. That is because Equations (2.58) and (2.59) are not pointwise but $\mathcal{L}^2(0,l)$ equalities, and should therefore be interpreted to mean

$$\left\| f(x) - \sum_{n=0}^{\infty} a_n \cos\frac{n\pi}{l}x \right\| = 0, \qquad \left\| f(x) - \sum_{n=0}^{\infty} b_n \sin\frac{n\pi}{l}x \right\| = 0,$$

respectively.

By Definition 1.29, the coefficients a_n and b_n are determined by the formulas

$$a_n = \frac{\langle f, \cos(n\pi x/l)\rangle}{\|\cos(n\pi x/l)\|^2}, \qquad n \in \mathbb{N}_0, \qquad b_n = \frac{\langle f, \sin(n\pi x/l)\rangle}{\|\sin(n\pi x/l)\|^2}, \qquad n \in \mathbb{N}.$$

Because

$$\left\|\cos\left(\frac{n\pi}{l}x\right)\right\|^2 = \int_0^l \cos^2\left(\frac{n\pi}{l}x\right)dx = \begin{cases} l, & n = 0 \\ l/2, & n \in \mathbb{N} \end{cases},$$
$$\left\|\sin\left(\frac{n\pi}{l}x\right)\right\|^2 = \int_0^l \sin^2\left(\frac{n\pi}{l}x\right)dx = \frac{l}{2}, \qquad n \in \mathbb{N},$$

the coefficients in the expansions (2.58) and (2.59) are therefore given by

$$a_0 = \frac{1}{l}\int_0^l f(x)dx, \tag{2.60}$$

$$a_n = \frac{2}{l}\int_0^l f(x)\cos\frac{n\pi}{l}x\,dx, \tag{2.61}$$

$$b_n = \frac{2}{l}\int_0^l f(x)\sin\frac{n\pi}{l}x\,dx. \tag{2.62}$$

Consider, for example, the constant function defined on $[0,l]$ by $f(x)=1$. For every $n\in\mathbb{N}$,

$$\langle 1, \sin(n\pi x/l)\rangle = \int_0^l \sin\frac{n\pi x}{l}dx = \frac{l}{n\pi}(1-\cos n\pi) = \frac{l}{n\pi}(1-(-1)^n).$$

Therefore $b_n = (2/n\pi)(1-(-1)^n)$ and we can write

$$\begin{aligned} 1 &= \frac{2}{\pi}\sum_{n=1}^{\infty}\left(\frac{1-(-1)^n}{n}\right)\sin\frac{n\pi}{l}x \\ &= \frac{4}{\pi}\left(\sin x + \frac{1}{3}\sin 3x + \frac{1}{5}\sin 5x + \cdots\right). \end{aligned}$$

As pointed out above, this equality should, of course, be understood to mean

$$\left\| 1 - \frac{2}{\pi}\sum_{k=1}^{\infty}\left(\frac{1-(-1)^k}{k}\right)\sin\frac{k\pi}{l}x\right\| = 0,$$

or

$$\frac{2}{\pi}\sum_{k=1}^{n}\left(\frac{1-(-1)^k}{k}\right)\sin\frac{k\pi}{l}x \xrightarrow{\mathcal{L}^2} 1 \text{ as } n\to\infty.$$

On the other hand, since

$$\begin{aligned} \langle 1, \cos(n\pi x/l)\rangle &= \int_0^l \cos(n\pi x/l)dx = \begin{cases} 0, & n\in\mathbb{N} \\ l, & n=0 \end{cases}, \\ \|1\|^2 &= l, \end{aligned}$$

the coefficients a_n in the expansion (2.58) all vanish except for the first, which is $a_0 = 1$, hence the series reduces to the single term 1. But this is to be expected because the function f coincides with the first element of the sequence

$$\left(\cos\frac{n\pi x}{l} : n\in\mathbb{N}_0\right) = \left(1, \cos\frac{\pi x}{l}, \cos\frac{2\pi x}{l}, \cdots\right).$$

By the same token, based on Theorem 2.29, the sequence of eigenfunctions for the SL problem discussed in Example 2.21 is complete in $\mathcal{L}^2_\rho(1,b)$, where

$\rho(x) = 1/x$. Hence any function $f \in \mathcal{L}^2_\rho(1, b)$ can be expanded in a series of the form

$$\sum_{n=1}^{\infty} b_n \sin\left(\frac{n\pi}{\log b} \log x\right),$$

where

$$b_n = \frac{\langle f, \sin(n\pi \log x/\log b)\rangle_\rho}{\|\sin(n\pi \log x/\log b)\|^2_\rho}.$$

2.4.3 The Singular SL Problem

We end this section with a brief word about the singular SL problem. In the equation

$$Lu + \lambda\rho u = (pu')' + ru + \lambda\rho u = 0, \qquad a < x < b,$$

where p is smooth and ρ is positive and continuous, we have so far assumed that p does not vanish on the closed and bounded interval $[a, b]$. Any relaxation of these conditions leads to a singular problem. In this book we consider singular SL problems which result from one or both of the following situations.

1. $p(x) = 0$ at $x = a$ and/or $x = b$.
2. The interval (a, b) is infinite.

In the first instance the expression $\rho p(f'g - fg')$ vanishes at the endpoint where $p = 0$, hence no boundary condition is required at that endpoint. In particular, if $p(a) = p(b) = 0$ and the limits of u at a and b exist, then the equation

$$\rho p(f'g - fg')\big|_a^b = 0 \tag{2.63}$$

is satisfied and L is self-adjoint. If (a, b) is infinite, then it would be necessary for $\sqrt{\rho}u(x)$ to tend to 0 as $|x| \to \infty$ in order for u to lie in $\mathcal{L}^2_\rho$. The conclusions of Theorem 2.29 remain valid under these conditions, although we do not prove this. For a more extensive treatment of the subject, the reader is referred to [15] and [18]. The latter, by Anton Zettl, gives an up-to-date account of the history and development of the singular problem and some recent results of a more general nature than we have discussed in this book. In particular, using the more general methods of Lebesgue measure and integration, the smoothness conditions imposed on the coefficients in the SL equation may be replaced by much weaker integrability conditions.

In the chapters below we consider the following typical examples of singular SL equations.

1. *Legendre's Equation*

$$(1-x^2)u'' - 2xu' + n(n+1)u = 0, \qquad -1 < x < 1,$$

where $\lambda = n(n+1)$ and the function $p(x) = 1 - x^2$ vanishes at the endpoints $x = \pm 1$, hence only the existence of $\lim u$ at $x = \pm 1$ is required.

2. *Hermite's Equation*

$$u'' - 2xu' + 2nu = 0, \qquad x \in \mathbb{R}.$$

After multiplication by e^{-x^2}, this equation is transformed to the standard self-adjoint form

$$e^{-x^2}u'' - 2xe^{-x^2}u' + 2ne^{-x^2}u = 0,$$

where $\lambda = 2n,\ p(x) = e^{-x^2}$, and $\rho(x) = e^{-x^2}$.

3. *Laguerre's Equation*

$$xu'' - (1-x)u' + nu = 0, \qquad x > 0,$$

which is transformed to the standard form

$$xe^{-x}u'' - (1-x)e^{-x}u' + ne^{-x}u = 0$$

by multiplication by e^{-x}. Here $p(x) = xe^{-x}$ vanishes at $x = 0$.

4. *Bessel's Equation*

$$xu'' + u' - \frac{n^2}{x}u + \lambda xu = 0, \qquad x > 0.$$

Here both functions $p(x) = x$ and $\rho(x) = x$ vanish at $x = 0$.

The solutions of these equations provide important examples of the so-called special functions of mathematical physics, which we study in Chapters 4 and 5. But before that, in Chapter 3, we use the regular SL problem to develop the $\mathcal{L}^2$ theory of Fourier series for the trigonometric functions. The singular problem allows us to generalize this theory to other orthogonal functions.

EXERCISES

2.23 Determine the eigenvalues and eigenfunctions of the boundary-value problem

$$u'' + \lambda u = 0, \qquad a \leq x \leq b,$$
$$u(a) = u(b) = 0.$$

2.24 Determine the eigenvalues and eigenfunctions of the equation $x^2u'' - xu' + \lambda u = 0$ on $(1, e)$ under the boundary conditions $u(1) = u(e) = 0$. Write the form of the orthogonality relation between the eigenfunctions.

2.25 Verify that $p(f'\bar{g} - f\bar{g}')|_a^b = 0$ if f and g satisfy the separated homogeneous boundary conditions (2.35) and (2.36).

2.26 Verify that $p(f'\bar{g} - f\bar{g}')|_a^b = 0$ if f and g satisfy the periodic conditions $u(a) = u(b),\ u'(a) = u'(b)$, provided $p(a) = p(b)$.

2.27 Which of the following boundary conditions make $p(f'g - fg')|_a^b = 0$?

(a) $p(x) = 1,\ a \leq x \leq b,\ u(a) = u(b),\ u'(a) = u'(b)$.

(b) $p(x) = x,\ 0 < a \leq x \leq b,\ u(a) = u'(b) = 0$.

(c) $p(x) = \sin x,\ 0 \leq x \leq \pi/2,\ u(0) = 1,\ u(\pi/2) = 0$.

(d) $p(x) = e^{-x},\ 0 < x < 1,\ u(0) = u(1),\ u'(0) = u'(1)$.

(e) $p(x) = x^2,\ 0 < x < b,\ u(0) = u'(b),\ u(b) = u'(0)$.

(f) $p(x) = x^2,\ -1 < x < 1,\ u(-1) = u(1),\ u'(-1) = u'(1)$.

2.28 Which boundary conditions in Exercise 2.27 define a singular SL problem?

2.29 Determine the eigenvalues and eigenfunctions of the problem

$$[(x+3)^2y']' + \lambda y = 0, \qquad -2 \leq x \leq 1,$$
$$y(-2) = y(1) = 0.$$

2.30 Let

$$u'' + \lambda u = 0, \qquad 0 \leq x \leq \pi,$$
$$u(0) = 0, \qquad 2u(\pi) - u'(\pi) = 0.$$

(a) Show how the positive eigenvalues λ may be determined. Are there any eigenvalues in $(-\infty, 0]$?

(b) What are the corresponding eigenfunctions?

2.31 Discuss the sequence of eigenvalues and eigenfunctions of the problem

$$\begin{aligned} u'' + \lambda u &= 0, \qquad & 0 \leq x \leq l, \\ u'(0) &= u(0), \qquad & u'(l) = 0. \end{aligned}$$

2.32 Let

$$\begin{gathered} (pu')' + ru + \lambda u = 0, \qquad a < x < b, \\ u(a) = u(b) = 0. \end{gathered}$$

If $p(x) \geq 0$ and $r(x) \leq c$, prove that $\lambda \geq -c$.

2.33 Using Equation (2.42), prove that the norm of the operator T is also given by

$$\|T\| = \sup_{\|u\|=1} |\langle Tu, u \rangle|,$$

where $u \in C([a,b])$. Hint: $|\langle Tu, u \rangle| \leq \|T\|$ by the CBS inequality and Equation (2.42). To prove the reverse inequality, first show that $2 \operatorname{Re} \langle Tu, v \rangle \leq (\|u\|^2 + \|v\|^2) \sup |\langle Tu, u \rangle|$, then set $v = Tu / \|Tu\|$.

3
Fourier Series

This chapter deals with the theory and applications of Fourier series, named after Joseph Fourier (1768–1830), the French physicist who developed the series in his investigation of the transfer of heat. His results were later refined by others, especially the German mathematician Gustav Lejeune Dirichlet (1805–1859), who made important contributions to the convergence properties of the series.

The first section presents the $\mathcal{L}^2$ theory, which is based on the results of the regular SL problem as developed in the previous chapter. We saw in Example 2.31 that, based on Theorem 2.29, both sets of orthogonal functions $\{\cos(n\pi x/l) : n \in \mathbb{N}_0\}$ and $\{\sin(n\pi x/l) : n \in \mathbb{N}\}$ span the space $\mathcal{L}^2(0,l)$. We now show that their union spans $\mathcal{L}^2(-l,l)$, and thereby arrive at the fundamental theorem of Fourier series.

In Section 3.2 we consider the pointwise theory and prove the basic result in this connection, which is Theorem 3.9, using the properties of the Dirichlet kernel. The third and final section of this chapter is devoted to applications to boundary-value problems.

3.1 Fourier Series in $\mathcal{L}^2$

A function $\varphi : [-l,l] \to \mathbb{C}$ is said to be *even* if

$$\varphi(-x) = \varphi(x) \quad \text{for all } x \in [-l,l],$$

and *odd* if

$$\varphi(-x) = -\varphi(x) \quad \text{for all } x \in [-l,l].$$

The same definition applies if the compact interval $[-l, l]$ is replaced by $(-l, l)$ or $(-\infty, \infty)$. If φ is integrable on $[-l, l]$, then it follows that

$$\int_{-l}^{l} \varphi(x)dx = 2\int_{0}^{l} \varphi(x)dx$$

if φ is even, and

$$\int_{-l}^{l} \varphi(x)dx = 0$$

if φ is odd. Clearly, the sum of two even (odd) functions is even (odd), and the product of two even or odd functions is even, whereas the product of an even function and an odd function is odd.

An arbitrary function f defined on $[-l, l]$ can always be represented as a sum of the form

$$f(x) = \frac{1}{2}[f(x) + f(-x)] + \frac{1}{2}[f(x) - f(-x)], \tag{3.1}$$

where $f_e(x) = \frac{1}{2}[f(x)+f(-x)]$ is an even function and $f_o(x) = \frac{1}{2}[f(x)-f(-x)]$ is an odd function. f_e and f_o are referred to as the *even* and the *odd parts* (or *components*) of f, respectively. To show that the representation (3.1) is unique, let

$$f(x) = \tilde{f}_e(x) + \tilde{f}_o(x) = f_e(x) + f_o(x), \tag{3.2}$$

where $\tilde{f}_e$ is even and $\tilde{f}_o$ is odd. Replacing x by $-x$ gives

$$f(-x) = \tilde{f}_e(x) - \tilde{f}_o(x) = f_e(x) - f_o(x). \tag{3.3}$$

Adding and subtracting (3.2) and (3.3) we conclude that $\tilde{f}_e = f_e$ and $\tilde{f}_o = f_o$.

Any linear combination of even (odd) functions is even (odd), therefore the even and odd functions in $\mathcal{L}^2(-l, l)$ clearly form two complementary subspaces which are orthogonal to each other, in the sense that if $\varphi, \psi \in \mathcal{L}^2(-l, l)$ with φ even and ψ odd, then φ and ψ are orthogonal to each other, for their inner product is

$$\langle \varphi, \psi \rangle = \int_{-l}^{l} \varphi(x)\psi(x)dx = 0$$

because the function $\varphi\psi$ is odd.

Let $f = f_e + f_o$ be any function in $\mathcal{L}^2(-l, l)$. Clearly both f_e and f_o belong to $\mathcal{L}^2(-l, l)$. The restriction of f_e to $(0, l)$ can be represented in $\mathcal{L}^2$ in terms of the (complete) orthogonal set $\{\cos(n\pi x/l) : n \in \mathbb{N}_0\}$ as

$$f_e(x) = a_0 + \sum_{n=1}^{\infty} a_n \cos\frac{n\pi}{l}x, \qquad 0 \le x \le l, \tag{3.4}$$

where, according to the formulas (2.60) and (2.61),

$$a_0 = \frac{1}{l}\int_0^l f_e(x)dx,$$
$$a_n = \frac{2}{l}\int_0^l f_e(x)\cos\frac{n\pi}{l}x\, dx, \qquad n \in \mathbb{N}.$$

The orthogonal set $\{\sin(n\pi x/l) : n \in \mathbb{N}\}$ also spans $\mathcal{L}^2(0,l)$, hence

$$f_o(x) = \sum_{n=1}^{\infty} b_n \sin\frac{n\pi}{l}x, \qquad 0 \le x \le l, \tag{3.5}$$

where

$$b_n = \frac{2}{l}\int_0^l f_o(x)\sin\frac{n\pi}{l}x\, dx, \qquad n \in \mathbb{N}.$$

Because both sides of (3.4) are even functions on $[-l,l]$, the equality extends to $[-l,l]$; and similarly the two sides of equation (3.5) are odd and this allows us to extend the equality to $[-l,l]$. Hence we have the following representation of any $f \in \mathcal{L}^2(-l,l)$,

$$\begin{aligned} f(x) &= a_0 + \sum_{n=1}^{\infty} a_n \cos\frac{n\pi}{l}x + \sum_{n=1}^{\infty} b_n \sin\frac{n\pi}{l}x \\ &= a_0 + \sum_{n=1}^{\infty}\left(a_n\cos\frac{n\pi}{l} + b_n\sin\frac{n\pi}{l}x\right), \qquad -l \le x \le l, \end{aligned} \tag{3.6}$$

the equality being, of course, in $\mathcal{L}^2(-l,l)$. On the other hand, using the properties of even and odd functions, we have, for all $n \in \mathbb{N}_0$,

$$\int_{-l}^{l} f_e(x)\sin\frac{n\pi}{l}x\, dx = \int_{-l}^{l} f_o(x)\cos\frac{n\pi}{l}x\, dx = 0,$$
$$\int_0^l f_e(x)\cos\frac{n\pi}{l}x\, dx = \frac{1}{2}\int_{-l}^{l} f_e(x)\cos\frac{n\pi}{l}x\, dx = \frac{1}{2}\int_{-l}^{l} f(x)\cos\frac{n\pi}{l}x\, dx,$$
$$\int_0^l f_o(x)\sin\frac{n\pi}{l}x\, dx = \frac{1}{2}\int_{-l}^{l} f_o(x)\sin\frac{n\pi}{l}x\, dx = \frac{1}{2}\int_{-l}^{l} f(x)\sin\frac{n\pi}{l}x\, dx.$$

Hence the coefficients a_n and b_n in the representation (3.6) are given by

$$a_0 = \frac{1}{2l}\int_{-l}^{l} f(x)dx,$$
$$a_n = \frac{1}{l}\int_{-l}^{l} f(x)\cos\frac{n\pi}{l}x\, dx, \qquad n \in \mathbb{N},$$
$$b_n = \frac{1}{l}\int_{-l}^{l} f(x)\sin\frac{n\pi}{l}x\, dx, \qquad n \in \mathbb{N}.$$

This result is confirmed by the following example.

Example 3.1

Find the eigenfunctions of the equation

$$u'' + \lambda u = 0, \qquad -l \leq x \leq l,$$

which satisfy the periodic boundary conditions

$$u(-l) = u(l), \qquad u'(-l) = u'(l).$$

Solution

This is an SL problem in which the boundary conditions are not separated but periodic. Because the operator $-d^2/dx^2$ is self-adjoint under these conditions (see Exercise 2.26), the conclusions of Theorem 2.29 still apply (except for the uniqueness of the eigenfunctions), as pointed out in Remark 2.30, hence its eigenfunctions are orthogonal and complete in $\mathcal{L}^2(-l, l)$. If $\lambda < 0$ it is straightforward to check that the differential equation has only the trivial solution under the given periodic boundary conditions. If $\lambda = 0$ the general solution is $u(x) = c_1 x + c_2$, and the boundary conditions yield the eigenfunction $u_0(x) = 1$.

For positive values of λ the solution of the equation is

$$c_1 \cos\sqrt{\lambda}x + c_2 \sin\sqrt{\lambda}x.$$

Applying the boundary conditions to this solution leads to the pair of equations

$$c_2 \sin\sqrt{\lambda}l = 0,$$
$$c_1\sqrt{\lambda}\sin\sqrt{\lambda}l = 0.$$

Because $\sqrt{\lambda} > 0$ and c_1 and c_2 cannot both be 0, the eigenvalues must satisfy $\sin\sqrt{\lambda}\, l = 0$. Thus the eigenvalues in $[0, \infty)$ are given by

$$\lambda_n = \left(\frac{n\pi}{l}\right)^2, \qquad n \in \mathbb{N}_0,$$

and the corresponding eigenfunctions are

$$1, \ \cos\frac{\pi}{l}x, \ \sin\frac{\pi}{l}x, \ \cos\frac{2\pi}{l}x, \ \sin\frac{2\pi}{l}x, \ \cos\frac{3\pi}{l}x, \ \sin\frac{3\pi}{l}x, \ \ldots .$$

Note that every eigenvalue $\lambda_n = (n\pi/l)^2$ is associated with two eigenfunctions, $\cos(n\pi x/l)$ and $\sin(n\pi x/l)$, except $\lambda_0 = 0$ which is associated with the single eigenfunction 1. This does not violate Theorem 2.29 which states that each eigenvalue corresponds to a single eigenfunction, for that part of the theorem only applies when the boundary conditions are separated. In the example at hand the boundary conditions yield the pair of equations

$$c_1\sqrt{\lambda}\sin\sqrt{\lambda}l = 0, \qquad c_2 \sin\sqrt{\lambda}l = 0,$$

which do not determine the constants c_1 and c_2 when $\lambda = (n\pi/l)^2$, so we can set $c_1 = 0$ to pick up the eigenfunction $\sin(n\pi x/l)$ and $c_2 = 0$ to pick up $\cos(n\pi x/l)$.

We have therefore proved the following.

Theorem 3.2 (Fundamental Theorem of Fourier Series)

The orthogonal set of functions

$$\{1, \cos\frac{n\pi}{l}x, \sin\frac{n\pi}{l}x : n \in \mathbb{N}\}$$

is complete in $\mathcal{L}^2(-l,l)$, in the sense that any function $f \in \mathcal{L}^2(-l,l)$ can be represented by the series

$$f(x) = a_0 + \sum_{n=1}^{\infty}\left(a_n\cos\frac{n\pi}{l}x + b_n\sin\frac{n\pi}{l}x\right), \qquad -l \le x \le l, \tag{3.7}$$

where

$$a_0 = \frac{\langle f, 1\rangle}{\|1\|^2} = \frac{1}{2l}\int_{-l}^{l} f(x)dx, \tag{3.8}$$

$$a_n = \frac{\langle f, \cos(n\pi x/l)\rangle}{\|\cos(n\pi x/l)\|^2} = \frac{1}{l}\int_{-l}^{l} f(x)\cos\frac{n\pi}{l}x\, dx, \qquad n \in \mathbb{N}, \tag{3.9}$$

$$b_n = \frac{\langle f, \sin(n\pi x/l)\rangle}{\|\sin(n\pi x/l)\|^2} = \frac{1}{l}\int_{-l}^{l} f(x)\sin\frac{n\pi}{l}x\, dx, \qquad n \in \mathbb{N}. \tag{3.10}$$

The right-hand side of Equation (3.7) is called the Fourier series expansion of f, and the coefficients a_n and b_n in the expansion are the Fourier coefficients of f.

Remark 3.3

1. If f is an even function, the coefficients $b_n = 0$ for all $n \in \mathbb{N}$ and f is then represented on $[-l, l]$ by a cosine series,

$$f(x) = a_0 + \sum_{n=1}^{\infty} a_n\cos\frac{n\pi}{l}x,$$

in which the coefficients are given by

$$a_0 = \frac{1}{l}\int_0^l f(x)dx, \qquad a_n = \frac{2}{l}\int_0^l f(x)\cos\frac{n\pi}{l}x\, dx, \qquad n \in \mathbb{N}.$$

Conversely, if f is odd, then $a_n = 0$ for all $n \in \mathbb{N}_0$ and f is represented by a sine series,

$$f(x) = \sum_{n=1}^{\infty} b_n \sin \frac{n\pi}{l} x,$$

where

$$b_n = \frac{2}{l} \int_0^l f(x) \sin \frac{n\pi}{l} x \, dx, \qquad n \in \mathbb{N}.$$

2. The equality expressed by (3.7) between f and its Fourier series is not necessarily pointwise, but in $\mathcal{L}^2(-l,l)$. It means that

$$\left\| f(x) - \left[a_0 + \sum_{n=1}^{N} \left(a_n \cos \frac{n\pi}{l} x + b_n \sin \frac{n\pi}{l} x \right) \right] \right\|^2 =$$

$$\|f\|^2 - l \left[2 |a_0|^2 + \sum_{n=1}^{N} (|a_n|^2 + |b_n|^2) \right] \to 0 \quad \text{as } N \to \infty.$$

Because $f \in \mathcal{L}^2(-l,l)$, this convergence clearly implies that the positive series $\sum |a_n|^2$ and $\sum |b_n|^2$ both converge and therefore $\lim a_n = \lim b_n = 0$.

3. If the Fourier series of f converges uniformly on $[-l,l]$, then, by Corollary 1.19, its sum is continuous and the equality (3.7) is pointwise provided f is continuous on $[-l,l]$.

Example 3.4

The function

$$f(x) = \begin{cases} -1, & -\pi < x < 0 \\ 0, & x = 0 \\ 1, & 0 < x \leq \pi, \end{cases}$$

shown in Figure 3.1,

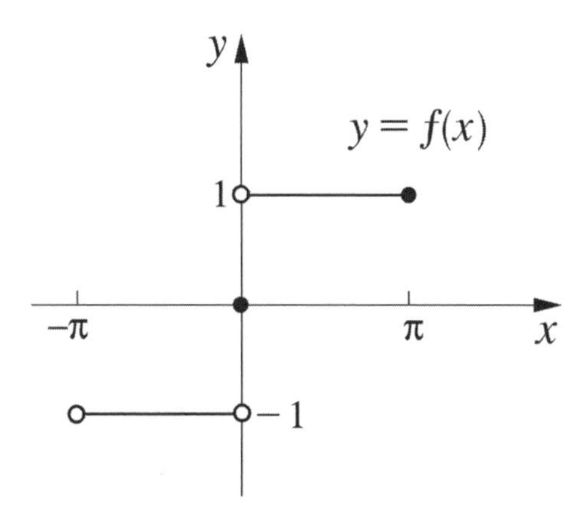

Figure 3.1

is clearly in $\mathcal{L}^2(-\pi,\pi)$. To obtain its Fourier series expansion, we first note that f is odd on $(-\pi,\pi)$, hence its Fourier series reduces to a sine series. The Fourier coefficients are given by

$$\begin{aligned} b_n &= \frac{1}{\pi}\int_{-\pi}^{\pi} f(x)\sin nx\,dx \\ &= \frac{2}{\pi}\int_0^{\pi} \sin nx\,dx \\ &= \frac{2}{n\pi}(1-\cos n\pi). \end{aligned}$$

b_n is therefore $4/n\pi$ when n is odd, and 0 when n is even, and tends to 0 as $n\to\infty$, in agreement with Remark 3.3 above. Thus we have

$$\begin{aligned} f(x) &= \sum_{n=1}^{\infty} b_n \sin nx \\ &= \frac{4}{\pi}\sin x + \frac{4}{3\pi}\sin 3x + \frac{4}{5\pi}\sin 5x + \cdots \\ &= \frac{4}{\pi}\sum_{n=0}^{\infty}\frac{1}{2n+1}\sin(2n+1)x. \end{aligned} \tag{3.11}$$

Observe how the terms of the partial sums

$$S_N(x) = \frac{4}{\pi}\sum_{n=0}^{N}\frac{1}{2n+1}\sin(2n+1)x$$

add up in Figure 3.2 to approximate the graph of f. Note also that the Fourier series of f is 0 at $\pm\pi$ whereas $f(\pi)=1$ and $f(-\pi)$ is not defined , which just says that Equation (3.11) does not hold at every point in $[-\pi,\pi]$. We shall have more to say about that in the next section.

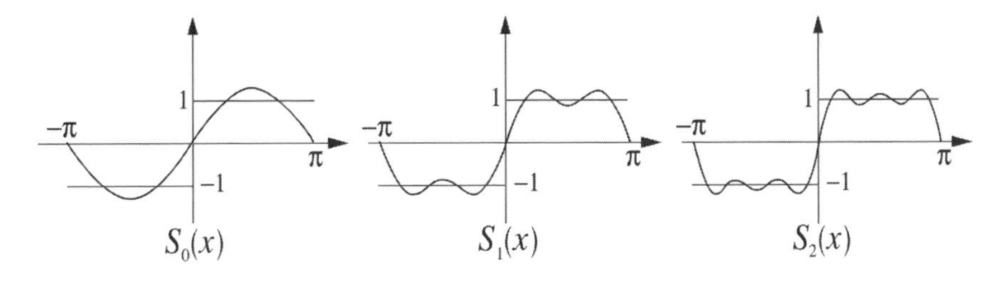

Figure 3.2 The sequence of partial sums S_N.

Theorem 3.2 applies to real as well as complex functions in $\mathcal{L}^2(-l,l)$. When f is real there is a clear advantage to using the orthogonal sequence of real trigonometric functions

$$1,\ \cos\frac{\pi}{l}x,\ \sin\frac{\pi}{l}x,\ \cos\frac{2\pi}{l}x,\ \sin\frac{2\pi}{l}x,\ \cdots$$

to expand f, for then the Fourier coefficients are also real. Alternatively, we could use Euler's relation

$$e^{ix} = \cos x + i\sin x, \qquad x \in \mathbb{R},$$

to express the Fourier series in exponential form. To that end we define the coefficients

$$\begin{aligned} c_0 &= a_0 = \frac{1}{2l}\int_{-l}^{l} f(x)dx, \\ c_n &= \frac{1}{2}(a_n - ib_n) = \frac{1}{2l}\int_{-l}^{l} f(x)e^{-in\pi x/l}dx, \\ c_{-n} &= \frac{1}{2}(a_n + ib_n) = \frac{1}{2l}\int_{-l}^{l} f(x)e^{in\pi x/l}dx, \qquad n \in \mathbb{N}, \end{aligned}$$

so that

$$\begin{aligned} a_0 &+ \sum_{n=1}^{N}\left(a_n\cos\frac{n\pi}{l}x + b_n\sin\frac{n\pi}{l}x\right) \\ &= c_0 + \sum_{n=1}^{N}\left[(c_n + c_{-n})\cos\frac{n\pi}{l}x + i(c_n - c_{-n})\sin\frac{n\pi}{l}x\right] \\ &= c_0 + \sum_{n=1}^{N}\left(c_n e^{in\pi x/l} + c_{-n}e^{-in\pi x/l}\right) = \sum_{n=-N}^{N} c_n e^{in\pi x/l}. \end{aligned}$$

Because $|c_n| \leq |a_n| + |b_n| \leq 2\,|c_n|$ for all $n \in \mathbb{Z}$, the two sides of this equation converge or diverge together as N increases. Thus, in the limit as $N \to \infty$, the right-hand side converges to the sum

$$\sum_{n=-\infty}^{\infty} c_n e^{in\pi x/l} \equiv c_0 + \sum_{n=1}^{\infty}(c_n e^{in\pi x/l} + c_{-n}e^{-in\pi x/l}),$$

and we obtain the exponential form of the Fourier series as follows.

Corollary 3.5

Any function $f \in \mathcal{L}^2(-l,l)$ can be represented by the Fourier series

$$f(x) = \sum_{n=-\infty}^{\infty} c_n e^{in\pi x/l}, \tag{3.12}$$

where

$$c_n = \frac{\langle f, e^{in\pi x/l}\rangle}{\left\|e^{in\pi x/l}\right\|^2} = \frac{1}{2l}\int_{-l}^{l} f(x)e^{-in\pi x/l}dx, \qquad n \in \mathbb{Z}. \tag{3.13}$$

We saw in Example 1.12 that the set $\{e^{inx} : n \in \mathbb{Z}\}$ is orthogonal in $\mathcal{L}^2(-\pi,\pi)$, hence $\{e^{in\pi x/l} : n \in \mathbb{Z}\}$ is orthogonal in $\mathcal{L}^2(-l,l)$. We can also show that $\{e^{in\pi x/l} : n \in \mathbb{Z}\}$ forms a complete set of eigenfunctions of the SL problem discussed in Example 3.1, where each eigenvalue $n^2\pi^2/l^2$ is associated with the pair of eigenfunctions $e^{\pm in\pi x/l}$, and thereby arrive at the representation (3.12) directly. In this representation, the Fourier coefficients c_n are defined by the single formula (3.13) instead of the three formulas for a_0, a_n, and b_n, which is a clear advantage. Using (3.12) and (3.13) to expand the function in Example 3.4, we obtain

$$\begin{aligned} c_n &= \frac{1}{2\pi}\int_{-\pi}^{\pi} f(x)e^{-inx}dx \\ &= \frac{1}{2\pi}\left[\int_0^{\pi} e^{-inx}dx - \int_{-\pi}^{0} e^{-inx}dx\right] \\ &= \frac{1}{i\pi}\int_0^{\pi} \sin nx\, dx \\ &= \frac{1}{in\pi}(1-\cos n\pi), \end{aligned}$$

$$\begin{aligned} f(x) &= \frac{1}{i\pi}\sum_{n=-\infty}^{\infty}\frac{1}{n}\left[1-(-1)^n\right]e^{inx} \\ &= \frac{2}{i\pi}\sum_{n=0}^{\infty}\frac{1}{2n+1}[e^{i(2n+1)x} - e^{-i(2n+1)}] \\ &= \frac{4}{\pi}\sum_{n=0}^{\infty}\frac{1}{2n+1}\sin(2n+1)x. \end{aligned}$$

EXERCISES

3.1 Verify that the set $\{e^{in\pi x/l} : n \in \mathbb{Z}\}$ is orthogonal in $\mathcal{L}^2(-l,l)$. What is the corresponding orthonormal set?

3.2 Is the series

$$\sum \frac{1}{2n+1}\sin(2n+1)x$$

uniformly convergent? Why?

3.3 Determine the Fourier series for the function f defined on $[-1,1]$ by

$$f(x)=\begin{cases}0, & -1\leq x<0\\ 1, & 0\leq x\leq 1.\end{cases}$$

3.4 Expand the function $f(x)=\pi-|x|$ in a Fourier series on $[-\pi,\pi]$, and prove that the series converges uniformly.

3.5 Expand the function $f(x)=x^2+x$ in a Fourier series on $[-2,2]$. Is the convergence uniform?

3.6 If the series $\sum|a_n|$ converges prove that the sum $\sum_{n=1}^{\infty}a_n\cos nx$ is finite for every x in the interval $[-\pi,\pi]$, where it represents a continuous function.

3.7 Prove that the real trigonometric series $\sum(a_n\cos nx+b_n\sin nx)$ converges in $\mathcal{L}^2(-\pi,\pi)$ if, and only if, the numerical series $\sum(a_n^2+b_n^2)$ converges.

3.2 Pointwise Convergence of Fourier Series

We say that a function $f:\mathbb{R}\to\mathbb{C}$ is *periodic* in p, where p is a positive number, if

$$f(x+p)=f(x)\quad\text{for all } x\in\mathbb{R},$$

and p is then called a *period* of f. When f is periodic in p it is also periodic in any positive multiple of p, because

$$\begin{aligned}f(x+np)&=f(x+(n-1)p+p)\\&=f(x+(n-1)p)\\&=\cdots\\&=f(x).\end{aligned}$$

It also follows that

$$f(x-np)=f(x-np+p)=\cdots=f(x),$$

hence the equality $f(x+np)=f(x)$ is true for any $n\in\mathbb{Z}$. The functions $\sin x$ and $\cos x$, for example, are periodic in 2π, whereas $\sin(ax)$ and $\cos(ax)$ are periodic in $2\pi/a$, where $a>0$. A constant function is periodic in any positive number.

A function f which is periodic in p and integrable on $[0,p]$ is clearly integrable over any finite interval and, furthermore, its integral has the same value over all intervals of length p. In other words

$$\int_x^{x+p} f(t)dt = \int_0^p f(t)dt \quad \text{for all } x \in \mathbb{R}. \tag{3.14}$$

This may be proved by using the periodicity of f to show that $\int_0^x f(t)dt = \int_p^{p+x} f(t)dt$ (Exercise 3.8).

When the trigonometric series

$$S_n(x) = a_0 + \sum_{k=1}^{n} (a_k \cos kx + b_k \sin kx)$$

converges on $\mathbb{R}$, its sum

$$S(x) = a_0 + \sum_{k=1}^{\infty} (a_k \cos kx + b_k \sin kx)$$

is clearly a periodic function in 2π, for 2π is a common period of all its terms. If a_k and b_k are the Fourier coefficients of some $\mathcal{L}^2(-\pi,\pi)$ function f, that is, if

$$\begin{aligned} a_0 &= \frac{1}{2\pi}\int_{-\pi}^{\pi} f(x)dx, \\ a_n &= \frac{1}{\pi}\int_{-\pi}^{\pi} f(x)\cos nx \, dx, \\ b_n &= \frac{1}{\pi}\int_{-\pi}^{\pi} f(x)\sin nx \, dx, \qquad n \in \mathbb{N}, \end{aligned}$$

then we know that S_n converges to f in $\mathcal{L}^2(-\pi,\pi)$,

$$S_n \xrightarrow{\mathcal{L}^2} f.$$

We now wish to investigate the conditions under which S_n converges pointwise to f on $[-\pi,\pi]$. More specifically, if f is extended from $[-\pi,\pi]$ to $\mathbb{R}$ as a periodic function by the equation

$$f(x+2\pi) = f(x) \quad \text{for all } x \in \mathbb{R},$$

when does the equality $f(x) = S(x)$ hold at every $x \in \mathbb{R}$? To answer this question we first need to introduce some definitions.

Definition 3.6

1. A function f defined on a bounded interval I, where $(a,b) \subseteq I \subseteq [a,b]$, is said to be *piecewise continuous* if

(i) f is continuous on (a,b) except for a finite number of points $\{x_1, x_2, \ldots, x_n\}$,

(ii) The right-hand and left-hand limits

$$\lim_{x\to x_i^+} f(x_i) = f(x_i^+), \qquad \lim_{x\to x_i^-} f(x) = f(x_i^-)$$

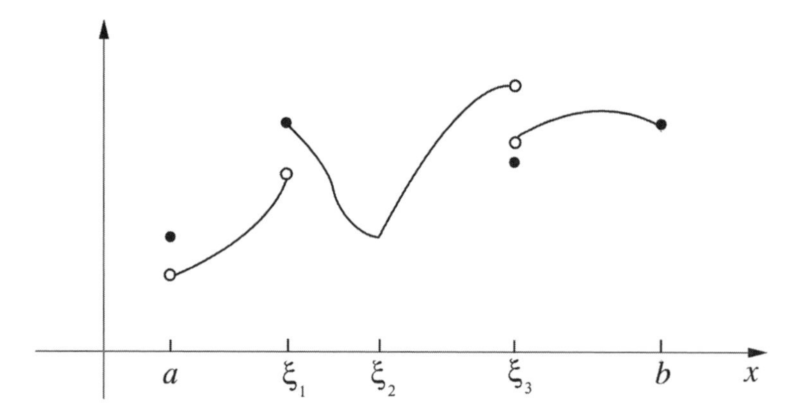

Figure 3.3 A piecewise smooth function.

exist at every point of discontinuity x_i, and

(iii) $\lim_{x \to a^+} f(x) = f(a^+)$ and $\lim_{x \to b^-} f(x) = f(b^-)$ exist at the endpoints.

2. f is *piecewise smooth* if f and f' are both piecewise continuous.

3. If the interval I is unbounded, then f is *piecewise continuous* (*smooth*) if it is piecewise continuous (smooth) on every bounded subinterval of I.

It is clear from the definition that a piecewise continuous function f on a bounded interval is a bounded function whose discontinuities are the result of finite "jumps" in its values. Its derivative f' is not defined at the points where f is discontinuous. f' is also not defined where it has a jump discontinuity (where the graph of f has a sharp "corner", as shown in Figure 3.3). Consequently, if f' is piecewise continuous on (a, b), there is a finite sequence of points (which includes the points of discontinuity of f)

$$a < \xi_1 < \xi_2 < \xi_3 < \cdots < \xi_m < b,$$

where f is not differentiable, but f' is continuous on each open subinterval (a, ξ_1), (ξ_i, ξ_{i+1}), and (ξ_m, b), $1 \leq i \leq m-1$.

Note that $f'(x^\pm)$ are the left-hand and right-hand limits of f' at x, given by

$$f'(x^-) = \lim_{x \to x^-} f'(x) = \lim_{h \to 0^+} \frac{f(x^-) - f(x-h)}{h},$$

$$f'(x^+) = \lim_{x \to x^+} f'(x) = \lim_{h \to 0^+} \frac{f(x+h) - f(x^+)}{h},$$

which are quite different from the left-hand and right-hand derivatives of f at x. The latter are determined by the limits

$$\lim_{h \to 0^\pm} \frac{f(x+h) - f(x)}{h},$$

which may not exist at ξ_i even if f is piecewise smooth. Conversely, not every differentiable function is piecewise smooth (see Exercises 3.10 and 3.11). A continuous function is obviously piecewise continuous, but it may not be piecewise smooth, such as

$$f(x) = \sqrt{x}, \qquad 0 \leq x \leq 1,$$

in as much as $\lim_{x\to 0^+} f'(x)$ does not exist.

The next two lemmas are used to prove Theorem 3.9, which is the main result of this section.

Lemma 3.7

If f is a piecewise continuous function on $[-\pi, \pi]$, then

$$\lim_{n\to\infty} \int_{-\pi}^{\pi} f(x) \cos nx dx = \lim_{n\to\infty} \int_{-\pi}^{\pi} f(x) \sin nx dx = \lim_{n\to\infty} \int_{-\pi}^{\pi} f(x) e^{\pm inx} dx = 0.$$

Proof

Suppose $x_1, \ldots, x_p$ are the points of discontinuity of f on $(-\pi, \pi)$ arranged in increasing order. Because $|f|$ is continuous and bounded on (x_k, x_{k+1}) for every $0 \leq k \leq p$, where $x_0 = -\pi$ and $x_{p+1} = \pi$, it is square integrable on all such intervals and

$$\int_{-\pi}^{\pi} |f(x)|^2 \, dx = \int_{-\pi}^{x_1} |f(x)|^2 \, dx + \cdots + \int_{x_p}^{\pi} |f(x)|^2 \, dx.$$

Consequently f belongs to $\mathcal{L}^2(-\pi, \pi)$. Its Fourier coefficients a_n, b_n, and c_n therefore tend to 0 as $n \to \infty$ (see Remarks 1.31 and 3.3). □

Lemma 3.8

For any real number $\alpha \neq 2m\pi$, $m \in \mathbb{Z}$,

$$\frac{1}{2} + \cos\alpha + \cos 2\alpha + \cdots + \cos n\alpha = \frac{\sin(n + \frac{1}{2})\alpha}{2 \sin \frac{1}{2}\alpha}. \tag{3.15}$$

Proof

Using the exponential expression for the cosine function, the left-hand side of (3.15) can be written in the form

$$\frac{1}{2} + \sum_{k=1}^{n} \cos k\alpha = \frac{1}{2} \sum_{k=-n}^{n} e^{ik\alpha}$$

$$= \frac{1}{2} e^{-in\alpha} \sum_{k=0}^{2n} e^{ik\alpha}$$
$$= \frac{1}{2} e^{-in\alpha} \sum_{k=0}^{2n} (e^{i\alpha})^k.$$

Because $e^{i\alpha} = 1$ if, and only if, α is an integral multiple of 2π,

$$\frac{1}{2} e^{-in\pi} \sum_{k=0}^{2n} (e^{i\alpha})^k = \frac{1}{2} e^{-in\alpha} \frac{1-(e^{i\alpha})^{2n+1}}{1-e^{i\alpha}}$$
$$= \frac{1}{2} \frac{e^{-in\alpha} - e^{i(n+1)\alpha}}{1-e^{i\alpha}}, \qquad \alpha \neq 0, \pm 2\pi, \dots . \tag{3.16}$$

Multiplying the numerator and denominator of this last expression by $e^{-i\alpha/2}$, we obtain the right-hand side of (3.15). □

The expression

$$D_n(\alpha) = \frac{1}{2\pi} \sum_{k=-n}^{n} e^{ik\alpha} = \frac{1}{2\pi} + \frac{1}{\pi} \sum_{k=1}^{n} \cos k\alpha,$$

known as the *Dirichlet kernel*, is a continuous function of α which is even and periodic in 2π (Figure 3.4). Based on Lemma 3.8, the Dirichlet kernel is also represented for all real values of α by

$$D_n(\alpha) = \begin{cases} \dfrac{\sin(n+\frac{1}{2})\alpha}{2\pi \sin \frac{1}{2}\alpha}, & \alpha \neq 0, 2\pi, \dots \\ \dfrac{2n+1}{2\pi}, & \alpha = 0, 2\pi, \dots . \end{cases}$$

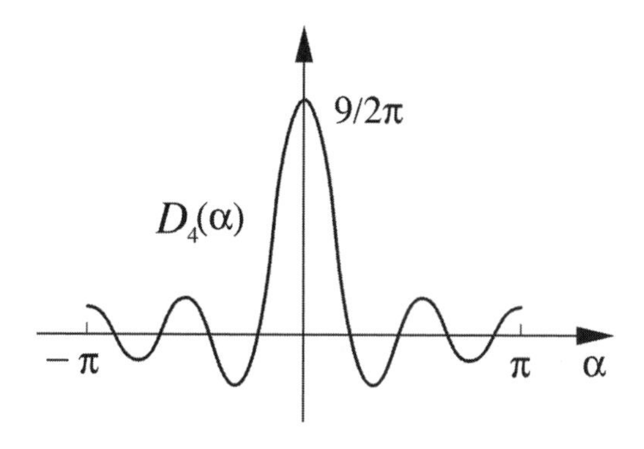

Figure 3.4 The Dirichlet kernel D_4.

Integrating $D_n(\alpha)$ on $0 \leq \alpha \leq \pi$, we obtain

$$\int_0^\pi D_n(\alpha)d\alpha = \frac{1}{\pi}\int_0^\pi \left(\frac{1}{2} + \sum_{k=1}^n \cos k\alpha\right) d\alpha = \frac{1}{2} \quad \text{for all } n \in \mathbb{N}. \tag{3.17}$$

Now we use Lemmas 3.7 and 3.8 to prove the following pointwise version of the fundamental theorem of Fourier series.

Theorem 3.9

Let f be a piecewise smooth function on $[-\pi, \pi]$ which is periodic in 2π. If

$$\begin{aligned} a_0 &= \frac{1}{2\pi}\int_{-\pi}^\pi f(x)dx, \\ a_n &= \frac{1}{\pi}\int_{-\pi}^\pi f(x)\cos nx \, dx, \\ b_n &= \frac{1}{\pi}\int_{-\pi}^\pi f(x)\sin nx \, dx, \quad n \in \mathbb{N}, \end{aligned}$$

then the Fourier series

$$S(x) = a_0 + \sum_{k=1}^\infty (a_n \cos nx + b_n \sin nx),$$

converges at every x in $\mathbb{R}$ to $\frac{1}{2}[f(x^+) + f(x^-)]$.

Proof

The nth partial sum of the Fourier series is

$$S_n(x) = a_0 + \sum_{k=1}^n (a_k \cos kx + b_n \sin kx).$$

Substituting the integral representations of the coefficients into this finite sum, and interchanging the order of integration and summation, we obtain

$$\begin{aligned} S_n(x) &= \frac{1}{\pi}\int_{-\pi}^\pi f(\xi)\left[\frac{1}{2} + \sum_{k=1}^n (\cos k\xi \cos kx + \sin k\xi \sin kx)\right] d\xi \\ &= \frac{1}{\pi}\int_{-\pi}^\pi f(\xi)\left[\frac{1}{2} + \sum_{k=1}^n \cos k(\xi - x)\right] d\xi. \end{aligned}$$

By the definition of D_n we can therefore write

$$\begin{aligned} S_n(x) &= \int_{-\pi}^\pi f(\xi)D_n(\xi - x)d\xi \\ &= \int_{-\pi-x}^{\pi-x} f(x+t)D_n(t)dt. \end{aligned}$$

As a functions of t, both $f(x+t)$ and $D_n(t)$ are periodic in 2π, so we can use the relation (3.14) to write

$$S_n(x) = \int_{-\pi}^{\pi} f(x+t)D_n(t)dt.$$

On the other hand, in view of (3.17) and the fact that D_n is an even function,

$$\frac{1}{2}[f(x^+) + f(x^-)] = f(x^-)\int_{-\pi}^{0} D_n(t)dt + f(x^+)\int_{0}^{\pi} D_n(t)dt.$$

Hence

$$\begin{aligned} S_n(x) - \frac{1}{2}[f(x^+) + f(x^-)] &= \int_{-\pi}^{0} [f(x+t) - f(x^-)]D_n(t)dt \\ &\quad + \int_{0}^{\pi} [f(x+t) - f(x^+)]D_n(t)dt, \end{aligned}$$

and we have to show that this sequence tends to 0 as $n \to \infty$.

Define the function

$$g(t) = \begin{cases} \dfrac{f(x+t) - f(x^-)}{e^{it} - 1}, & -\pi < t < 0 \\ \dfrac{f(x+t) - f(x^+)}{e^{it} - 1}, & 0 < t < \pi. \end{cases}$$

Using (3.16) we can write

$$\begin{aligned} S_n(x) - \frac{1}{2}[f(x^+) + f(x^-)] &= \int_{-\pi}^{\pi} g(t)(e^{it} - 1)D_n(t)dt \\ &= \frac{1}{2\pi}\int_{-\pi}^{\pi} g(t)(e^{i(n+1)t} - e^{-int})dt, \end{aligned}$$

and it remains to show that the right-hand side tends to 0 as $n \to \infty$. Because f is piecewise smooth on $[-\pi, \pi]$, the function g is also piecewise smooth on $[-\pi, \pi]$ except possibly at $t = 0$. At $t = 0$ we can use L'Hôpital's rule to obtain

$$\lim_{t \to 0^\pm} g(t) = \lim_{t \to 0^\pm} \frac{f'(x+t)}{ie^{it}} = -if'(x^\pm).$$

Consequently g is piecewise continuous on $[-\pi, \pi]$, its Fourier coefficients converge to 0 by Lemma 3.7, and therefore

$$\frac{1}{2\pi}\int_{-\pi}^{\pi} g(t)e^{\pm int}dt \to 0 \quad \text{as } n \to \infty.$$

□

Remark 3.10

1. If f is continuous on $[-\pi, \pi]$ then $\frac{1}{2}[f(x^+) + f(x^-)] = f(x)$ for every $x \in \mathbb{R}$, so that the Fourier series converges pointwise to f on $\mathbb{R}$. Otherwise, where f is discontinuous at x, its Fourier series converges to the median of the "jump" at x, which is $\frac{1}{2}[f(x^+) + f(x^-)]$, regardless of how $f(x)$ is defined. This is consistent with the observation that, because the Fourier series S is completely determined by its coefficients a_n and b_n, and because these coefficients are defined by integrals involving f, the coefficients would not be sensitive to a change in the value of $f(x)$ at isolated points. If f is defined at its points of discontinuity by the formula

$$f(x) = \frac{1}{2}[f(x^+) + f(x^-)],$$

then we would again have the pointwise equality $S(x) = f(x)$ on the whole of $\mathbb{R}$.

2. Using the exponential form of the Fourier series, we have, for any function f which satisfies the hypothesis of the theorem,

$$\lim_{n\to\infty} \sum_{k=-n}^{n} c_n e^{inx} = \frac{1}{2}[f(x^+) + f(x^-)] \quad \text{for all } x \in \mathbb{R},$$

where

$$c_n = \frac{1}{2\pi} \int_{\pi}^{\pi} f(x) e^{-inx} dx.$$

3. Theorem 3.9 gives sufficient conditions for the pointwise convergence of the Fourier series of f to $\frac{1}{2}[f(x^+) + f(x^-)]$. For an example of a function which is not piecewise smooth, but which can still be represented by a Fourier series, see page 91 [16], or Exercise 3.26.

As one would expect, the interval $[-\pi, \pi]$ may be replaced by any other interval which is symmetric with respect to the origin.

Corollary 3.11

Let f be a piecewise smooth function on $[-l, l]$ which is periodic in $2l$. If

$$a_0 = \frac{1}{2l} \int_{-l}^{l} f(x) dx,$$

$$a_n = \frac{1}{l} \int_{-l}^{l} f(x) \cos \frac{n\pi}{l} x \, dx,$$

$$b_n = \frac{1}{l} \int_{-l}^{l} f(x) \sin \frac{n\pi}{l} x \, dx, \qquad n \in \mathbb{N},$$

then the Fourier series

$$a_0 + \sum_{n=1}^{\infty} \left(a_n \cos \frac{n\pi}{l} x + b_n \sin \frac{n\pi}{l} x \right)$$

converges at every x in $\mathbb{R}$ to $\frac{1}{2}[f(x^+) + f(x^-)]$.

Proof

Setting $g(x) = f(lx/\pi)$ we see that $g(x + 2\pi) = g(x)$, and that g satisfies the conditions of Theorem 3.9. □

In Example 3.4 the function f is piecewise continuous on $(-\pi, \pi]$, hence its periodic extension to $\mathbb{R}$ satisfies the conditions of Theorem 3.9. Note that f is discontinuous at all integral multiples of π, where it has a jump discontinuity of magnitude $|f(x_n^+) - f(x_n^-)| = 2$. At $x = 0$ we have

$$\frac{1}{2}[f(0^+) + f(0^-)] = \frac{1}{2}[1 - 1] = 0,$$

which agrees with the value of its Fourier series

$$S(x) = \frac{4}{\pi} \sum_{n=0}^{\infty} \frac{1}{2n+1} \sin(2n+1)x$$

at $x = 0$. Because $f(0)$ was defined to be 0, the Fourier series converges to f at this point and, by periodicity, at all other even integral multiples of π. The same cannot be said of the point $x = \pi$, where $f(\pi) = 1 \neq S(\pi) = 0$. By periodicity the Fourier series does not converge to f at $x = \pm\pi, \pm 3\pi, \ldots$. To obtain convergence at these points (and hence at all points in $\mathbb{R}$), we would have to redefine f at π to be

$$f(\pi) = \frac{1}{2}[f(\pi^+) + f(\pi^-)] = 0.$$

In this same example, because f is continuous at $x = \pi/2$, we have $f(\pi/2) = S(\pi/2) = 1$. Hence

$$\begin{aligned} 1 &= \frac{4}{\pi} \sum_{n=0}^{\infty} \frac{1}{2n+1} \sin(2n+1)\frac{\pi}{2} \\ &= \frac{4}{\pi} \sum_{n=0}^{\infty} \frac{1}{2n+1} (-1)^n, \end{aligned}$$

which yields the following series representation of π :

$$\pi = 4\left(1 - \frac{1}{3} + \frac{1}{5} - \frac{1}{7} + \cdots\right). \tag{3.18}$$

Given any real function f, we use the symbol f^+ to denote the positive part of f, that is,

$$\begin{aligned} f^+(x) &= \begin{cases} f(x) & \text{if } f(x) \geq 0 \\ 0 & \text{if } f(x) < 0 \end{cases} \\ &= \frac{1}{2}[f(x) + |f(x)|]. \end{aligned}$$

Similarly, the negative part of f is

$$f^-(x) = \frac{1}{2}[|f(x)| - f(x)] = \begin{cases} -f(x) & \text{if } f(x) \leq 0 \\ 0 & \text{if } f(x) > 0 \end{cases},$$

and we clearly have $f = f^+ - f^-$.

Example 3.12

The function $f(t) = (2\cos 100\pi t)^+$, shown graphically in Figure 3.5, describes the current flow, as a function of time t, through an electric semiconductor, also known as a half-wave rectifier. The current amplitude is 2 and its frequency is 50.

To expand f, which is clearly piecewise smooth, in a Fourier series we first have to determine its period. This can be done by setting $\pi/l = 100\pi$, from which we conclude that $l = 1/100$. Because f is an even function, $b_n = 0$ for all $n \in \mathbb{N}$ and a_n are given by

$$\begin{aligned} a_0 &= \frac{1}{2l}\int_{-l}^{l} f(t)dt \\ &= \frac{1}{l}\int_0^l f(t)dt \\ &= 100\int_0^{1/200} 2\cos 100\pi t dt \\ &= 2/\pi, \end{aligned}$$

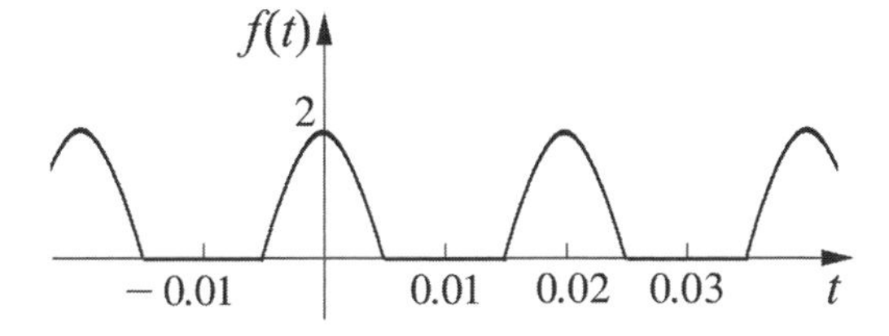

Figure 3.5 Rectified wave.

$$\begin{aligned}
a_n &= \frac{2}{l}\int_0^l f(t)\cos\frac{n\pi}{l}t\,dt \\
&= 200\int_0^{1/200} 2\cos 100\pi t\cos 100n\pi t\,dt \\
&= 200\int_0^{1/200} [\cos(n+1)100\pi t + \cos(n-1)100\pi t]dt, \qquad n\in\mathbb{N}.
\end{aligned}$$

When $n = 1$,

$$a_1 = 200\int_0^{1/200}(\cos 200\pi t + 1)dt = 1.$$

For all $n \geq 2$, we have

$$\begin{aligned}
a_n &= \frac{2}{\pi}\left[\frac{1}{n+1}\sin(n+1)100\pi t + \frac{1}{n-1}\sin(n-1)100\pi t\right]\Big|_0^{1/200} \\
&= \frac{2}{\pi}\left[\frac{1}{n+1}\sin(n+1)\frac{\pi}{2} + \frac{1}{n-1}\sin(n-1)\frac{\pi}{2}\right] \\
&= \frac{2}{\pi}\left(\frac{1}{n+1} - \frac{1}{n-1}\right)\cos n\frac{\pi}{2} \\
&= -\frac{4}{\pi(n^2-1)}\cos n\frac{\pi}{2}.
\end{aligned}$$

Thus

$$a_2 = \frac{4}{3\pi}, \qquad a_3 = 0, \qquad a_4 = -\frac{4}{15\pi}, \qquad a_5 = 0, \; \ldots \; .$$

Because f is a continuous function we finally obtain

$$\begin{aligned}
f(t) &= \frac{2}{\pi} + \cos 100\pi t + \frac{4}{\pi}\left(\frac{1}{3}\cos 200\pi t - \frac{1}{15}\cos 400\pi t + \cdots\right) \\
&= \frac{2}{\pi} + \cos 100\pi t - \frac{4}{\pi}\sum_{n=1}^{\infty}\frac{(-1)^n}{4n^2-1}\cos 200n\pi t \qquad \text{for all } t\in\mathbb{R}.
\end{aligned}$$

The continuity of f is consistent with the fact that the series converges uniformly (by the M-test). The converse is not true in general, for a convergent series can have a continuous sum without the convergence being uniform, as in the pointwise convergence of $\sum x^n$ to $1/(1-x)$ on $(-1,1)$. But this cannot happen with a Fourier series, as we now show.

Theorem 3.13

If f is a continuous function on the interval $[-\pi,\pi]$ such that $f(-\pi) = f(\pi)$ and f' is piecewise continuous on $(-\pi,\pi)$, then the series

$$\sum_{n=1}^{\infty}\sqrt{|a_n|^2 + |b_n|^2}$$

is convergent, where a_n and b_n are the Fourier coefficients of f defined by

$$a_n = \frac{1}{\pi}\int_{-\pi}^{\pi} f(x)\cos nx\,dx, \quad b_n = \frac{1}{\pi}\int_{-\pi}^{\pi} f(x)\sin nx\,dx.$$

Observe that the conditions imposed on f in this theorem are the same as those of Theorem 3.9 with piecewise continuity replaced by continuity on $[-\pi, \pi]$.

Proof

Because f' is piecewise continuous on $[-\pi, \pi]$, it belongs to $\mathcal{L}^2(-\pi, \pi)$ and its Fourier coefficients

$$\begin{aligned}
a_0' &= \frac{1}{2\pi}\int_{-\pi}^{\pi} f'(x)dx, \\
a_n' &= \frac{1}{\pi}\int_{-\pi}^{\pi} f'(x)\cos nx\,dx, \\
b_n' &= \frac{1}{\pi}\int_{-\pi}^{\pi} f'(x)\sin nx\,dx, \qquad n \in \mathbb{N},
\end{aligned}$$

exist. Integrating directly, or by parts, and using the relation $f(-\pi) = f(\pi)$, we obtain

$$\begin{aligned}
a_0' &= \frac{1}{2\pi}[f(\pi) - f(-\pi)] = 0, \\
a_n' &= \frac{1}{\pi} f(x)\cos nx\Big|_{-\pi}^{\pi} + \frac{n}{\pi}\int_{-\pi}^{\pi} f(x)\sin nx\,dx = nb_n, \\
b_n' &= \frac{1}{\pi} f(x)\sin nx\Big|_{-\pi}^{\pi} - \frac{n}{\pi}\int_{-\pi}^{\pi} f(x)\cos nx\,dx = -na_n.
\end{aligned}$$

Consequently,

$$\begin{aligned}
S_N &= \sum_{n=1}^{N} \sqrt{|a_n|^2 + |b_n|^2} \\
&= \sum_{n=1}^{N} \frac{1}{n}\sqrt{|a_n'|^2 + |b_n'|^2} \\
&\leq \left[\sum_{n=1}^{N} \frac{1}{n^2} \cdot \sum_{n=1}^{N} \left(|a_n'|^2 + |b_n'|^2\right)\right]^{1/2},
\end{aligned}$$

where we use the CBS inequality in the last relation. By Bessel's inequality (1.22) we have

$$\sum_{n=1}^{N}(|a_n'|^2 + |b_n'|^2) \leq \int_{-\pi}^{\pi} |f'(x)|^2 \, dx < \infty.$$

The series $\sum 1/n^2$ converges, and thus the increasing sequence S_N is bounded above and is therefore convergent. □

Corollary 3.14

If f satisfies the conditions of Theorem 3.13, then the convergence of the Fourier series

$$a_0 + \sum_{n=1}^{\infty} (a_n \cos nx + b_n \sin nx) \tag{3.19}$$

to f on $[-\pi, \pi]$ is uniform and absolute.

Proof

The extension of f from $[-\pi, \pi]$ to $\mathbb{R}$ by the periodicity relation

$$f(x + 2\pi) = f(x) \quad \text{for all } x \in \mathbb{R}$$

is a continuous function which satisfies the conditions of Theorem 3.9, hence the Fourier series (3.19) converges to $f(x)$ at every $x \in \mathbb{R}$. But

$$|a_n \cos nx + b_n \sin nx| \leq |a_n| + |b_n| \leq 2\sqrt{|a_n|^2 + |b_n|^2},$$

and because $\sum \sqrt{|a_n|^2 + |b_n|^2}$ converges, the series (3.19) converges uniformly and absolutely by the M-test. □

Based on Corollaries 1.19 and 3.14, we can now assert that a function which meets the conditions of Theorem 3.9 is continuous if, and only if, its Fourier series is uniformly convergent on $\mathbb{R}$. Needless to say, this result remains valid if the period 2π is replaced by any other period $2l$.

Corollary 3.15

If f is piecewise smooth on $[-l, l]$ and periodic in $2l$, its Fourier series is uniformly convergent if, and only if, f is continuous.

EXERCISES

3.8 Prove Equation (3.14).

3.9 Determine which of the following functions are piecewise continuous and which are piecewise smooth.

(a) $f(x) = \sqrt[3]{x}$, $x \in \mathbb{R}$.

(b) $f(x) = |\sin x|$, $x \in \mathbb{R}$.

(c) $f(x) = \sqrt{x}$, $0 \leq x < 1$, $f(x+1) = f(x)$ for all $x \geq 0$.

(d) $f(x) = |x|^{3/2}$, $-1 \leq x \leq 1$, $f(x+2) = f(x)$ for all $x \in \mathbb{R}$.

(e) $f(x) = [x] = n$ for all $x \in [n, n+1)$, $n \in \mathbb{Z}$.

3.10 Show that the function $f(x) = |x|$ is piecewise smooth on $\mathbb{R}$, but not differentiable at $x = 0$. Give an example of a function which is piecewise smooth on $\mathbb{R}$ but not differentiable at any integer.

3.11 Show that the function

$$f(x) = \begin{cases} x^2 \sin(1/x), & x \neq 0 \\ 0, & x = 0, \end{cases}$$

is differentiable, but not piecewise smooth, on $\mathbb{R}$.

3.12 Assuming f is a piecewise smooth function on $(-l, l)$ which is periodic in $2l$, show that f is piecewise smooth on $\mathbb{R}$.

3.13 Suppose that the functions f and g are piecewise smooth on the interval I. Prove that their sum $f + g$ and product $f \cdot g$ are also piecewise smooth on I. What can be said about the quotient f/g?

3.14 Determine the zeros of the Dirichlet kernel D_n and its maximum value.

3.15 Verify that each of the following functions is piecewise smooth and determine its Fourier series.

(a) $f(x) = x$, $-\pi \leq x \leq \pi$, $f(x + 2\pi) = f(x)$ for all $x \in \mathbb{R}$.

(b) $f(x) = |x|$, $-1 \leq x \leq 1$, $f(x+2) = f(x)$ for all $x \in \mathbb{R}$.

(c) $f(x) = \sin^2 x$, $x \in \mathbb{R}$.

(d) $f(x) = \cos^3 x$, $x \in \mathbb{R}$.

(e) $f(x) = e^x$, $-2 \leq x \leq 2$, $f(x+4) = f(x)$ for all $x \in \mathbb{R}$.

(f) $f(x) = x^3$, $-l \leq x \leq l$, $f(x + 2l) = f(x)$ for all $x \in \mathbb{R}$.

3.16 In Exercise 3.15 determine where the convergence of the Fourier series is uniform.

3.17 Determine the sum $S(x)$ of the Fourier series at $x = \pm 2$ in Exercise 3.15(e) and at $x = \pm l$ in Exercise 3.15(f).

3.18 Use the Fourier series expansion of $f(x) = x, \ -\pi < x \leq \pi, \ f(x + 2\pi) = f(x)$ on $\mathbb{R}$, to obtain Equation (3.18).

3.19 Use the Fourier series expansion of $f(x) = |x|, \ -\pi \leq x \leq \pi, \ f(x + 2\pi) = f(x)$ on $\mathbb{R}$, to obtain a representation of π^2.

3.20 Use the Fourier series expansion of f in Example 3.12 to obtain a representation of π.

3.21 Determine the Fourier series expansion of $f(x) = x^2$ on $[-\pi, \pi]$, then use the result to evaluate each of the following series.

(a) $\sum_{n=1}^{\infty} \frac{1}{n^2}$.

(b) $\sum_{n=1}^{\infty} (-1)^n \frac{1}{n^2}$.

(c) $\sum_{n=1}^{\infty} \frac{1}{(2n)^2}$.

(d) $\sum_{n=1}^{\infty} \frac{1}{(2n-1)^2}$.

3.22 Expand the function $f(x) = |\sin x|$ in a Fourier series on $\mathbb{R}$. Verify that the series is uniformly convergent, and use the result to evaluate the series $\sum_{n=1}^{\infty}(4n^2 - 1)^{-1}$ and $\sum_{n=1}^{\infty}(-1)^{n+1}(4n^2 - 1)^{-1}$.

3.23 Show that the function $f(x) = |\sin x|$ has a piecewise smooth derivative. Expand f' in a Fourier series and determine the value of the series at $x = n\pi$ and at $x = \pi/2 + n\pi$, $n \in \mathbb{Z}$. Sketch f and f'.

3.24 Suppose f is a piecewise smooth function on $[0, l]$. The even periodic extension of f is the function $\tilde{f}_e$ defined on $\mathbb{R}$ by

$$\tilde{f}_e(x) = \begin{cases} f(x), & 0 \leq x \leq l \\ f(-x), & -l < x < 0, \end{cases}$$

$$\tilde{f}(x + 2l) = \tilde{f}(x), \ x \in \mathbb{R};$$

and the odd periodic extension of f is

$$\tilde{f}_o(x) = \begin{cases} f(x), & 0 \leq x \leq l \\ -f(-x), & -l < x < 0, \end{cases}$$

$$\tilde{f}(x + 2l) = \tilde{f}(x), \ x \in \mathbb{R}.$$

(a) Show that, if f is continuous on $[0, l]$, then $\tilde{f}_e$ is continuous on $\mathbb{R}$, but that $\tilde{f}_o$ is continuous if, and only if, $f(0) = f(l) = 0$.

(b) Obtain the Fourier series expansions of $\tilde{f}_e$ and $\tilde{f}_0$.

(c) Given $f(x) = x$ on $[0, 1]$, determine the Fourier series of $\tilde{f}_e$ and $\tilde{f}_o$.

3.25 For each of the following functions $f : \mathbb{R} \to \mathbb{R}$, determine the value of the Fourier series at $x = 0$ and $x = \pi$, and compare the result with $f(0)$ and $f(\pi)$:

(a) Even periodic extension of $\sin x$ from $[0, \pi]$ to $\mathbb{R}$.

(b) Odd periodic extension of $\cos x$ from $[0, \pi]$ to $\mathbb{R}$.

(c) Odd periodic extension of e^x from $[0, \pi]$ to $\mathbb{R}$.

3.26 Show that the Fourier series of $f(x) = x^{1/3}$ converges at every point in $[-\pi, \pi]$. Because f is continuous and lies in $\mathcal{L}^2(-\pi, \pi)$, conclude that its Fourier series converges pointwise to $f(x)$ at every $x \in (-\pi, \pi)$, although f is not piecewise smooth on $(-\pi, \pi)$.

3.3 Boundary-Value Problems

Fourier series play a crucial role in the construction of solutions to boundary-value and initial-value problems for partial differential equations. A solution of a partial differential equation will naturally be a function of more than one variable. When the equation is linear and homogeneous, an effective method for obtaining its solution is by separation of variables. This is based on the assumption that a solution, say $u(x, y)$, can be expressed as a product of a function of x and a function of y,

$$u(x, y) = v(x)w(y). \tag{3.20}$$

After substituting into the partial differential equation, if we can arrange the terms in the resulting equation so that one side involves only the variable x and the other only y, then each side must be a constant. This gives rise to two ordinary differential equations for the functions v and w which are, presumably, easier to solve than the original equation. The loss of generality involved in the initial assumption (3.20) is compensated for by taking linear combinations of all such solutions $v(x)w(y)$ over the permissible values of the parametric constant. Two well-known examples of boundary-value problems which describe physical phenomena are now given and solved using separation of variables.

3.3.1 The Heat Equation

Consider the heat equation in one space dimension

$$\frac{\partial u}{\partial t} = k\frac{\partial^2 u}{\partial x^2}, \qquad 0 < x < l,\ t > 0, \tag{3.21}$$

which governs the heat flow along a thin bar of length l, where $u(x,t)$ is the temperature of the bar at point x and time t, and k is a positive physical constant which depends on the material of the bar. To determine u we need to know the boundary conditions at the ends of the bar and the initial distribution of temperature along the bar. Suppose the boundary conditions on u are simply

$$u(0,t) = u(l,t) = 0, \qquad t > 0; \tag{3.22}$$

that is, the bar ends are held at 0 temperature, and the initial temperature distribution along the bar is

$$u(x,0) = f(x), \qquad 0 < x < l, \tag{3.23}$$

for some given function f. We wish to determine $u(x,t)$ for all points (x,t) in the strip $(0,l) \times (0,\infty)$.

Let

$$u(x,t) = v(x)w(t).$$

Substituting into (3.21), we obtain

$$v(x)w'(t) = kv''(x)w(t),$$

which, after dividing by kvw, becomes

$$\frac{v''(x)}{v(x)} = \frac{w'(t)}{kw(t)}. \tag{3.24}$$

Equation (3.24) cannot hold over the strip $(0,l) \times (0,\infty)$ unless each side is a constant, say $-\lambda^2$. The reason for choosing the constant to be $-\lambda^2$ instead of λ is clarified in Remark 3.16. Thus (3.24) breaks down into two ordinary differential equations,

$$v'' + \lambda^2 v = 0, \tag{3.25}$$

$$w' + \lambda^2 kw = 0. \tag{3.26}$$

The solutions of (3.25) and (3.26), are

$$v(x) = a\cos\lambda x + b\sin\lambda x, \tag{3.27}$$

$$w(t) = ce^{-\lambda^2 kt}, \tag{3.28}$$

where a, b, and c are constants of integration. Because these solutions depend on the parameter λ, we indicate this dependence by writing v_λ and w_λ. Thus the solutions we seek have the form

$$u_\lambda(x,t) = (a_\lambda \cos \lambda x + b_\lambda \sin \lambda x)e^{-\lambda^2 kt}, \tag{3.29}$$

where a_λ and b_λ are arbitrary constants which also depend on λ.

The first boundary condition in (3.22) implies

$$u_\lambda(0,t) = a_\lambda e^{-\lambda^2 kt} = 0 \quad \text{for all } t > 0,$$

from which we conclude that $a_\lambda = 0$. The second boundary condition gives

$$u_\lambda(l,t) = b_\lambda \sin \lambda l \; e^{-\lambda^2 kt} \quad \text{for all } t > 0,$$

hence $b_\lambda \sin \lambda l = 0$. Because we cannot allow b_λ to be 0, or we get the trivial solution $u_\lambda \equiv 0$ which cannot satisfy the initial condition (unless $f \equiv 0$), we conclude that

$$\sin \lambda l = 0,$$

and therefore λl is an integral multiple of π. Thus the parameter λ can only assume the discrete values

$$\lambda_n = \frac{n\pi}{l}, \qquad n \in \mathbb{N}.$$

Notice that we have retained only the positive values of n, because the negative values have the effect of changing the sign of $\sin \lambda_n x$, and can therefore be accommodated by the constants $b_{\lambda_n} = b_n$. The case $n = 0$ yields the trivial solution. Thus we arrive at the sequence of solutions to the heat equation, all satisfying the boundary conditions (3.22),

$$u_n(x,t) = b_n \sin \frac{n\pi}{l} x \; e^{-(n\pi/l)^2 kt}, \qquad n \in \mathbb{N}.$$

Equations (3.21) and (3.22) are linear and homogeneous, therefore the formal sum

$$u(x,t) = \sum_{n=1}^{\infty} u_n(x,t) = \sum_{n=1}^{\infty} b_n \sin \frac{n\pi}{l} x \; e^{-(n\pi/l)^2 kt}, \tag{3.30}$$

defines a more general solution of these equations (by the superposition principle).

Now we apply the initial condition (3.23) to the expression (3.30):

$$u(x,0) = \sum_{n=1}^{\infty} b_n \sin \frac{n\pi}{l} x = f(x).$$

Assuming that the function f is piecewise smooth on $[0, l]$, its odd extension into $[-l, l]$ satisfies the conditions of Corollary 3.11 in the interval $[-l, l]$, and the coefficients b_n must therefore be the Fourier coefficients of f; that is,

$$b_n = \frac{2}{l} \int_0^l f(x) \sin \frac{n\pi}{l} x \, dx.$$

Here we are tacitly assuming that $f(x) = \frac{1}{2}[f(x^+) + f(x^-)]$ for all x. This completely determines $u(x, t)$, represented by the series (3.30), as the solution of the system of Equations (3.21) through (3.23) in $[0, l] \times [0, \infty)$.

Remark 3.16

1. Although the assumption $u(x, t) = v(x)w(t)$ at the outset places a restriction on the form of the solutions (3.29) that we obtain, the linear combination (3.30) of such solutions restores the generality that was lost, in as much as the sequence $\sin(n\pi x/l)$ spans $\mathcal{L}^2(0, l)$.

2. The reason we chose the constant $v''(x)/v(x) = w'(t)/kw(t)$ to be negative is that a positive constant would change the solutions of Equation (3.25) from trigonometric to hyperbolic functions, and these cannot be used to expand $f(x)$ (unless λ is imaginary). The square on λ was introduced merely for the sake of convenience, in order to avoid using the square root sign in the representation of the solutions.

In higher space dimensions, the homogeneous heat equation takes the form

$$u_t = k\Delta u, \tag{3.31}$$

where Δ is the Laplacian operator in space variables. In $\mathbb{R}^n$ the Laplacian is given in Cartesian coordinates by

$$\Delta = \frac{\partial^2}{\partial x_1^2} + \frac{\partial^2}{\partial x_2^2} + \cdots + \frac{\partial^2}{\partial x_n^2}.$$

Example 3.17

The dynamic (i.e., time-dependent) temperature distribution on a rectangular conducting sheet of length a and width b, whose boundary is held at 0 temperature, is described by the following boundary-value problem for the heat equation in $\mathbb{R}^2$.

$$u_t = k(u_{xx} + u_{yy}), \qquad (x, y) \in (0, a) \times (0, b), \ t > 0,$$

$$u(0, y, t) = u(a, y, t) = 0, \qquad y \in (0, b), \ t > 0,$$

$$u(x,0,t) = u(x,b,t) = 0, \qquad x \in (0,a),\ t > 0,$$
$$u(x,y,0) = f(x,y), \qquad (x,y) \in (0,a) \times (0,b),$$

where we use the more compact notation

$$u_t = \frac{\partial u}{\partial t}, \quad u_x = \frac{\partial u}{\partial x}, \quad u_{xx} = \frac{\partial^2 u}{\partial x^2}, \quad u_{xy} = \frac{\partial^2 u}{\partial y \partial x}, \quad u_{yy} = \frac{\partial^2 u}{\partial y^2}, \quad \ldots .$$

The assumption that

$$u(x,y,t) = v(x,y)w(t)$$

leads to the equation $v(x,y)w'(t) = kw(t)[v_{xx}(x,y) + v_{yy}(x,y)]$. After dividing by vw and noting that the left-hand side of the resulting equation depends on t and the right-hand side depends on (x,y), we again assume a separation constant $-\lambda^2$ and obtain the pair of equations

$$w' + \lambda^2 kw = 0, \tag{3.32}$$
$$v_{xx} + v_{yy} + \lambda^2 v = 0. \tag{3.33}$$

Equation (3.32), as is (3.26), is solved by the exponential function (3.28). We can use separation of variables again to solve (3.33). Substituting $v(x,y) = X(x)Y(y)$ into (3.33) leads to

$$\frac{X''(x)}{X(x)} + \frac{Y''(y)}{Y(y)} + \lambda^2 = 0,$$

where the variables are separable,

$$\frac{X''(x)}{X(x)} = -\frac{Y''(y)}{Y(y)} - \lambda^2 = -\mu^2,$$

and $-\mu^2$ is the second separation constant. The resulting pair of equations has the general solutions

$$X(x) = A \cos \mu x + B \sin \mu x,$$
$$Y(y) = A' \cos \sqrt{\lambda^2 - \mu^2} y + B' \sin \sqrt{\lambda^2 - \mu^2} y.$$

The boundary conditions at $x = 0$ and $y = 0$ imply $A = A' = 0$. From the condition at $x = a$ we obtain $\mu = \mu_n = n\pi/a$ for all positive integers n, and from the condition at $y = b$ we conclude that

$$\sqrt{\lambda^2 - \mu_n^2} = m\pi/b, \qquad m \in \mathbb{N},$$

and therefore

$$\lambda = \lambda_{mn} = \sqrt{\frac{n^2}{a^2} + \frac{m^2}{b^2}}\pi.$$

Thus we arrive at the double sequence of functions

$$u_{mn}(x,y,t) = b_{mn}e^{-k\lambda_{mn}^2 t}\sin\frac{n\pi}{a}x\,\sin\frac{m\pi}{b}y, \qquad m,n\in\mathbb{N},$$

which satisfy the heat equation and the boundary conditions.

Before applying the initial condition we form the formal double sum

$$u(x,y,t) = \sum_{m,n=1}^{\infty} b_{mn}e^{-k\lambda_{mn}^2 t}\sin\frac{n\pi}{a}x\,\sin\frac{m\pi}{b}y, \tag{3.34}$$

which satisfies the boundary conditions on the sides of the rectangular sheet. Now the condition at $t=0$ implies

$$f(x,y) = \sum_{m,n=1}^{\infty} b_{mn}\sin\frac{n\pi}{a}x\,\sin\frac{m\pi}{b}y. \tag{3.35}$$

Assuming f is square integrable over the rectangle $(0,a)\times(0,b)$, we first extend f as an odd function of x and y into $(-a,a)\times(-b,b)$. Then we multiply Equation (3.35) by $\sin(n'\pi/a)x$ and integrate over $(-a,a)$ to obtain

$$\sum_{n=1}^{\infty} b_{mn'}\sin\frac{m\pi}{b}y\left\|\sin\frac{n'\pi}{a}x\right\|^2 = 2\int_0^a f(x,y)\sin\frac{n'\pi}{a}x\,dx.$$

Multiplying by $\sin(m'\pi/b)y$ and integrating over $(-b,b)$ yields

$$b_{m'n'}\left\|\sin\frac{n'\pi}{a}x\right\|^2\left\|\sin\frac{m'\pi}{b}y\right\|^2 = 4\int_0^b\int_0^a f(x,y)\sin\frac{n'\pi}{a}x\,\sin\frac{m'\pi}{b}y\,dxdy.$$

Since

$$\|\sin(n'\pi/a)x\|^2 = \int_{-a}^a \sin^2\left(\frac{n'\pi}{a}x\right)dx = a$$

and

$$\|\sin(m'\pi/b)y\|^2 = \int_{-b}^b \sin^2\left(\frac{m'\pi}{a}y\right)dy = b,$$

the coefficients in (3.34) are given by

$$b_{mn} = \frac{4}{ab}\int_0^a\int_0^b f(x,y)\sin\frac{n\pi}{a}x\,\sin\frac{m\pi}{b}y\,dxdy. \tag{3.36}$$

The expression on the right-hand side of (3.35) is called a double Fourier series of f and the equality holds in $\mathcal{L}^2$ over the rectangle $(0,a)\times(0,b)$. Under appropriate smoothness conditions on the function f, such as the continuity of f and its partial derivatives on $[0,a]\times[0,b]$, (3.35) becomes a pointwise equality in which the Fourier coefficients b_{mn} are defined by (3.36). The series representation (3.34) is then convergent and satisfies the heat equation in the rectangle $(0,a)\times(0,b)$, the conditions on its boundary, and the initial condition at $t=0$.

It is worth noting that the presence of the exponential function in (3.30) and (3.34) means that the solution of the heat equation decays exponentially to 0 as $t\to\infty$. This is to be expected from a physical point of view, because heat flows from points of higher temperature to points of lower temperature; and because the edges of the domain in each case are held at (absolute) zero, all heat eventually seeps out. Thus, in the limit as $t\to\infty$, the heat equation becomes

$$\Delta u = 0,$$

known as Laplace's equation. This is a static equation which, in this case, describes the steady state distribution of temperature over a domain in the absence of heat sources; but, more generally, Laplace's equation governs many other physical fields which can be described by a potential function, such as gravitational, electrostatic, and fluid velocity fields. Some boundary-value problems for the Laplace equation are given in Exercises 3.34 through 3.37. We shall return to this equation in the next chapter.

3.3.2 The Wave Equation

If a thin, flexible, and weightless string, which is stretched between two fixed points, say $x=0$ and $x=l$, is given a small vertical displacement and then released from rest, it will vibrate along its length in such a way that the (vertical) displacement at point x and time t, denoted by $u(x,t)$, satisfies the wave equation in one space dimension,

$$u_{tt} = c^2 u_{xx}, \qquad 0<x<l,\ t>0, \tag{3.37}$$

c being a positive constant determined by the material of the string. This equation describes the transverse vibrations of an ideal string, and it differs from the heat equation only in the fact that the time derivative is of second, instead of first, order; but, as we now show, the solutions are very different. The boundary conditions on u are naturally

$$u(0,t) = u(l,t) = 0, \qquad t>0, \tag{3.38}$$

and the initial conditions are

$$u(x,0) = f(x), \qquad u_t(x,0) = 0, \qquad 0 < x < l, \tag{3.39}$$

$f(x)$ being the initial shape of the string and 0 the initial velocity. Here two initial conditions are needed because the time derivative in the wave equation is of second order.

Again using separation of variables, with $u(x,t) = v(x)w(t)$, leads to

$$\frac{v''(x)}{v(x)} = \frac{w''(t)}{c^2 w(t)} = -\lambda^2.$$

Hence

$$\begin{aligned} v''(x) + \lambda^2 v(x) &= 0, \qquad 0 < x < l, \\ w''(t) + c^2\lambda^2 w(t) &= 0, \qquad t > 0, \end{aligned}$$

whose solutions are

$$\begin{aligned} v(x) &= a\cos\lambda x + b\sin\lambda x, \\ w(t) &= a'\cos c\lambda t + b'\sin c\lambda t. \end{aligned}$$

As in the case of the heat equation, the boundary conditions imply $a = 0$ and $\lambda = n\pi/l$ with $n \in \mathbb{N}$. Thus each function in the sequence

$$u_n(x,t) = \left(a_n \cos\frac{cn\pi}{l}t + b_n \sin\frac{cn\pi}{l}t\right)\sin\frac{n\pi}{l}x, \qquad n \in \mathbb{N},$$

satisfies the wave equation and the boundary conditions at $x = 0$ and $x = l$, so (by superposition) the same is true of the (formal) sum

$$u(x,t) = \sum_{n=1}^{\infty}\left(a_n \cos\frac{cn\pi}{l}t + b_n \sin\frac{cn\pi}{l}t\right)\sin\frac{n\pi}{l}x. \tag{3.40}$$

The first initial condition implies

$$\sum_{n=1}^{\infty} a_n \sin\frac{n\pi}{l}x = f(x), \qquad 0 < x < l.$$

Assuming f is piecewise smooth, we again invoke Corollary 3.11 to conclude that

$$a_n = \frac{2}{l}\int_0^l f(x)\sin\frac{n\pi}{l}x\,dx. \tag{3.41}$$

The second initial condition gives

$$\sum_{n=1}^{\infty} \frac{cn\pi}{l} b_n \cos\frac{cn\pi}{l}t = 0, \qquad 0 < x < l,$$

hence $b_n = 0$ for all n.

The solution of the wave equation (3.37) under the given boundary and initial conditions (3.38) and (3.39) is therefore

$$u(x,t) = \sum_{n=1}^{\infty} a_n \cos \frac{cn\pi}{l} t \, \sin \frac{n\pi}{l} x, \tag{3.42}$$

where a_n are determined by the Fourier coefficient formula (3.41). Had the string been released with an initial velocity described, for example, by the function

$$u_t(x,0) = g(x),$$

the coefficients b_n in (3.40) would then be determined by the initial condition

$$\sum_{n=1}^{\infty} \frac{cn\pi}{l} b_n \sin \frac{cn\pi}{l} x = g(x)$$

as

$$b_n = \frac{2}{cn\pi} \int_0^l g(x) \sin \frac{cn\pi}{l} x \, dx.$$

Note how the vibrations of the string, as described by the series (3.42), continue indefinitely as $t \to \infty$. That is because the wave equation (3.37) does not take into account any energy loss, such as may be due to air resistance or internal friction in the string. This is in contrast to the solution (3.30) of the heat equation which tends to 0 as $t \to \infty$ as explained above. In Exercise 3.32, where the wave equation includes a friction term due to air resistance, the situation is different.

EXERCISES

3.27 Use separation of variables to solve each of the following boundary-value problems for the heat equation.

(a) $u_t = u_{xx}, \; 0 < x < \pi, \; t > 0,$

$u(0,t) = u(\pi,t) = 0, \; t > 0,$

$u(x,0) = \sin^3 x, \; 0 < x < \pi.$

(b) $u_t = k u_{xx}, \; 0 < x < 3, \; t > 0,$

$u(0,t) = u_x(3,t) = 0, \; t > 0,$

$u(x,0) = \sin \frac{\pi}{2} x - \sin \frac{5\pi}{6} x.$

The condition that $u_x = 0$ at an endpoint of the bar means that the endpoint is insulated, so that there is no heat flow through it.

3.28 If a bar of length l is held at a constant temperature T_0 at one end, and at T_1 at the other, the boundary-value problem becomes

$$\begin{aligned} u_t &= ku_{xx}, && 0 < x < l,\ t > 0, \\ u(0,t) &= T_0, && u(l,t) = T_1,\ \ t > 0, \\ u(x,0) &= f(x), && 0 < x < l. \end{aligned}$$

Solve this system of equations for $u(x,t)$. Hint: Assume that $u(x,t) = v(x,t) + \psi(x)$, where v satisfies the heat equation and the homogeneous boundary conditions $v(0,t) = v(l,t) = 0$.

3.29 Solve the following boundary-value problem for the wave equation.

$$\begin{aligned} u_{tt} &= u_{xx}, && 0 < x < \pi,\ l > 0, \\ u(0,t) &= u(l,t) = 0, && t > 0, \\ u(x,0) &= x(l-x), && u_t(x,0) = 0,\ \ 0 < x < l. \end{aligned}$$

3.30 Determine the vibrations of the string discussed in Section 3.3.2 if it is given an initial velocity $g(x)$ by solving

$$\begin{aligned} u_{tt} &= c^2 u_{xx}, && 0 < x < l,\ t > 0, \\ u(0,t) &= u(l,t) = 0, && t > 0, \\ u(x,0) &= f(x), && u_t(x,0) = g(x),\ \ 0 < x < l. \end{aligned}$$

Show that the solution can be represented by d'Alembert's formula

$$u(x,t) = \frac{1}{2}[\tilde{f}(x+ct) + \tilde{f}(x-ct)] + \frac{1}{2c}\int_{x-ct}^{x+ct} \tilde{g}(\tau)d\tau,$$

where $\tilde{f}$ and $\tilde{g}$ are the odd periodic extensions of f and g, respectively, from $(0,l)$ to $\mathbb{R}$.

3.31 If gravity is taken into account, the vibrations of a string stretched between $x = 0$ and $x = l$ are governed by the equation $u_{tt} = c^2 u_{xx} - g$, where g is the gravitational acceleration constant. Determine these vibrations under the homogeneous boundary conditions $u(0,t) = u(l,t) = 0$, and the homogeneous initial conditions $u(x,0) = u_t(x,0) = 0$.

3.32 If the air resistance to a vibrating string is proportional to its velocity, the resulting damped wave equation is

$$u_{tt} + ku_t = c^2 u_{xx}$$

for some constant k. Assuming that the initial shape of the string is $u(x,0) = 1$, $0 < x < 10$, and that it is released from rest, determine its subsequent shape for all $(x,t) \in (0,10) \times (0,\infty)$ under homogeneous boundary conditions.

3.33 The vibrations of a rectangular membrane which is fixed along its boundary are described by the system of equations

$$\begin{aligned} u_{tt} &= c^2(u_{xx} + u_{yy}), \qquad 0 < x < a, \qquad 0 < t < b,\ t > 0, \\ u(0,y,t) &= u(a,y,t) = 0, \qquad 0 < y < b, \qquad t > 0, \\ u(x,0,t) &= u(x,b,t) = 0, \qquad 0 < x < a, \qquad t > 0, \\ u(x,y,0) &= f(x,y),\ u_t(x,y,0) = g(x,y), \qquad 0 < x < a,\ \ 0 < y < b. \end{aligned}$$

Solve this system of equations in the rectangle $(0,a) \times (0,b)$ for all $t > 0$.

3.34 Determine the solution of Laplace's equation in the rectangle $(0,a) \times (0,b)$ under the conditions $u(0,y) = u(a,y) = 0$ on $0 < y < b$ and $u(x,0) = f(x)$, $u_y(x,b) = g(x)$ on $0 < x < a$.

3.35 Solve the boundary-value problem

$$\begin{aligned} u_{xx} + u_{yy} &= 0, \qquad 0 < x < 1, \qquad 0 < y < 1, \\ u(0,y) &= u_x(1,y) = 0, \qquad 0 < y < 1, \\ u(x,0) &= 0, \qquad u(x,1) = \sin \frac{3\pi}{2} x. \end{aligned}$$

3.36 Solve the boundary-value problem

$$\begin{aligned} u_{xx} + u_{yy} &= 0, \qquad 0 < x < \pi, \\ u_x(0,y) &= u_x(\pi,y) = 0, \qquad 0 < y < \pi, \\ u_y(x,0) &= \cos x,\ u_y(x,\pi) = 0, \qquad 0 < x < \pi. \end{aligned}$$

3.37 Laplace's equation in polar coordinates (r,θ) is given by

$$u_{rr} + \frac{1}{r} u_r + \frac{1}{r^2} u_{\theta\theta} = 0.$$

(a) Use separation of variables to show that the solutions of this equation in $\mathbb{R}^2$ are given by

$$u_n(r,\theta) = \begin{cases} a_0 + d_0 \log r, & n = 0 \\ (r^n + d_n r^{-n})(a_n \cos n\theta + b_n \sin n\theta), & n \in \mathbb{N}, \end{cases}$$

where a_n, b_n, and d_n are integration constants. Hint: Use the fact that $u(r,\theta + 2\pi) = u(r,\theta)$.

(b) Form the general solution $u(r,\theta) = \sum_{n=0}^{\infty} u_n(r,\theta)$ of Laplace's equation, then show that the solution of the equation inside the circle $r = R$ which satisfies $u(R,\theta) = f(\theta)$ is given by

$$u(r,\theta) = A_0 + \sum_{n=1}^{\infty} \left(\frac{r}{R}\right)^n (A_n \cos n\theta + B_n \sin n\theta),$$

for all $r \geq R,\ 0 \leq \theta < 2\pi$, where A_n and B_n are the Fourier coefficients of f.

(c) Determine the solution of Laplace's equation in polar coordinates outside the circle $r = R$ under the same boundary condition $u(R,\theta) = f(\theta)$.

4
Orthogonal Polynomials

In this chapter we consider three typical examples of singular SL problems whose eigenfunctions are real polynomials. Each set of eigenfunctions is generated by a particular choice of the coefficients in the eigenvalue equation

$$(pu')' + ru + \lambda \rho u = 0, \tag{4.1}$$

and of the interval $a < x < b$. In all cases the equation

$$p(u'v - uv')\big|_a^b = 0, \tag{4.2}$$

of course, has to be satisfied by any pair of eigenfunctions u and v.

If $\{\varphi_n : n \in \mathbb{N}_0\}$ is a complete set of eigenfunctions of some singular SL problem, then, being orthogonal and complete in $\mathcal{L}_\rho^2(a,b)$, the sequence (φ_n) may be used to expand any function $f \in \mathcal{L}_\rho^2(a,b)$ by the formula

$$f(x) = \sum_{n=0}^{\infty} \frac{\langle f, \varphi_n \rangle_\rho}{\|\varphi_n\|_\rho^2} \varphi_n(x) \tag{4.3}$$

in much the same way that the trigonometric functions $\cos nx$ or $\sin nx$ were used to represent a function in $\mathcal{L}^2(0,\pi)$. Thus we arrive at a generalization of the Fourier series which was associated with the choice $p(x) = 1$, $r(x) = 0$, $\rho(x) = 1$, and $(a,b) = (0,\pi)$ in (4.1). Consequently, the right-hand side of Equation (4.3) is referred to as a *generalized Fourier series* of f, and

$$c_n = \frac{\langle f, \varphi_n \rangle_\rho}{\|\varphi_n\|_\rho^2}, \qquad n \in \mathbb{N}_0,$$

are the *generalized Fourier coefficients* of f.

A corresponding result to Theorem 3.9 also applies to the generalized Fourier series: If f is piecewise smooth on (a, b), and

$$c_n = \frac{1}{\|\varphi_n\|_\rho^2} \int_a^b f(x)\varphi_n(x)\rho(x)dx,$$

then the series

$$S(x) = \sum_{n=0}^{\infty} c_n \varphi_n(x)$$

converges at every $x \in (a, b)$ to $\frac{1}{2}[f(x^+) + f(x^-)]$. A general proof of this result is beyond the scope of this treatment, but some relevant references to this topic may be found in [5]. Of course the periodicity property in $\mathbb{R}$, which was peculiar to the trigonometric functions, would not be expected to hold for a general orthogonal basis.

4.1 Legendre Polynomials

The equation

$$(1 - x^2)u'' - 2xu' + \lambda u = 0, \qquad -1 < x < 1, \tag{4.4}$$

is called Legendre's equation, after the French mathematician A.M. Legendre (1752–1833). It is one of the simplest examples of a singular SL problem, the singularity being due to the fact that $p(x) = 1 - x^2$ vanishes at the endpoints $x = \pm 1$. By rewriting Equation (4.4) in the equivalent form

$$u'' - \frac{2x}{1 - x^2}u' + \frac{\lambda}{1 - x^2}u = 0, \tag{4.5}$$

we see that the coefficients are analytic in the interval $(-1, 1)$, and the solution of the equation can therefore be represented by a power series about the point $x = 0$. Setting

$$u(x) = \sum_{k=0}^{\infty} c_k x^k, \qquad -1 < x < 1, \tag{4.6}$$

and substituting into Legendre's equation, we obtain

$$(1 - x^2)\sum_{k=2}^{\infty} k(k-1)c_k x^{k-2} - 2\sum_{k=1}^{\infty} kc_k x^k + \lambda \sum_{k=0}^{\infty} c_k x^k = 0$$

$$\sum_{k=0}^{\infty} [(k+2)(k+1)c_{k+2} + (\lambda - k^2 - k)c_k]x^k = 0,$$

from which

$$c_{k+2} = \frac{k(k+1) - \lambda}{(k+1)(k+2)} c_k. \tag{4.7}$$

Equation (4.7) is a *recursion formula* for the coefficients of the power series (4.6) which expresses the constants c_k for all $k \geq 2$ in terms of c_0 and c_1, and yields two independent power series solutions, one in even powers of x, and the other in odd powers.

If $\lambda = n(n+1)$, where $n \in \mathbb{N}_0$, the recursion relation (4.7) implies

$$0 = c_{n+2} = c_{n+4} = c_{n+6} = \cdots,$$

and it then follows that one of the two solutions is a polynomial. In that case (4.7) takes the form

$$c_{k+2} = \frac{k(k+1) - n(n+1)}{(k+1)(k+2)} c_k = \frac{(k-n)(k+n+1)}{(k+1)(k+2)} c_k. \tag{4.8}$$

Thus, with c_0 and c_1 arbitrary, we have

$$\begin{aligned}
c_2 &= -\frac{n(n+1)}{2!} c_0, & c_3 &= -\frac{(n-1)(n+2)}{3!} c_1 \\
c_4 &= \frac{n(n-2)(n+1)(n+3)}{4!} c_0, & c_5 &= \frac{(n-3)(n-1)(n+2)(n+4)}{5!} c_1 \\
&\ \vdots & &\ \vdots
\end{aligned}$$

and the solution of Legendre's equation takes the form

$$\begin{aligned}
u(x) &= c_0 \left[1 - \frac{n(n+1)}{2!} x^2 + \frac{n(n-2)(n+1)(n+3)}{4!} x^4 + \cdots \right] \\
&\quad + c_1 \left[x - \frac{(n-1)(n+2)}{3!} x^3 + \frac{(n-3)(n-1)(n+2)(n+4)}{5!} x^5 + \cdots \right] \\
&= c_0 u_0(x) + c_1 u_1(x),
\end{aligned}$$

where the power series $u_0(x)$ and $u_1(x)$ converge in $(-1, 1)$ and are linearly independent, the first being an even function and the second an odd function.

For each $n \in \mathbb{N}_0$ we therefore obtain a pair of linearly independent solutions,

$$\begin{aligned}
n = 0, \quad & u_0(x) = 1 \\
& u_1(x) = x + \frac{1}{3} x^3 + \frac{1}{5} x^5 + \cdots, \\
n = 1, \quad & u_0(x) = 1 - x^2 - \frac{1}{3} x^4 + \cdots \\
& u_1(x) = x,
\end{aligned}$$

$$\begin{aligned}
n = 2, \quad & u_0(x) = 1 - 3x^2 \\
& u_1(x) = x - \frac{2}{3}x^3 - \frac{1}{5}x^5 + \cdots, \\
n = 3, \quad & u_0(x) = 1 - 6x^2 + 3x^4 + \cdots \\
& u_1(x) = x - \frac{5}{3}x^3, \\
\vdots \quad & \quad \vdots,
\end{aligned}$$

one of which is a polynomial, and the other an infinite power series which converges in $(-1,1)$. We are mainly interested in the polynomial solution, which can be written in the form

$$a_n x^n + a_{n-2}x^{n-2} + a_{n-4}x^{n-4} + \cdots. \tag{4.9}$$

This is a polynomial of degree n which is either even or odd, depending on the integer n. By defining the coefficient of the highest power in the polynomial to be

$$a_n = \frac{(2n)!}{2^n(n!)^2} = \frac{1 \cdot 3 \cdot 5 \cdot \cdots \cdot (2n-1)}{n!}, \tag{4.10}$$

the resulting expression is called Legendre's polynomial of degree n, and is denoted by $P_n(x)$. As a result of this choice, we show in the next section that $P_n(1) = 1$ for all n. The other coefficients in (4.9) are determined in accordance with the recursion relation (4.8):

$$\begin{aligned}
a_{n-2} &= -\frac{n(n-1)}{2(2n-1)}a_n \\
&= -\frac{n(n-1)}{2(2n-1)}\frac{(2n)!}{2^n(n!)^2} \\
&= -\frac{(2n-2)!}{2^n(n-1)!(n-2)!} \\
a_{n-4} &= -\frac{(n-2)(n-3)}{4(2n-3)}a_{n-2} \\
&= \frac{(2n-4)!}{2^n 2!(n-2)!(n-4)!} \\
&\vdots \\
a_{n-2k} &= (-1)^k \frac{(2n-2k)!}{2^n k!(n-k)!(n-2k)!}, \qquad n \geq 2k,
\end{aligned}$$

where we arrive at the last equality by induction on k. The last coefficient in P_n is given by

$$a_0 = (-1)^{n/2}\frac{n!}{2^n(n/2)!(n/2)!}$$

if n is even, and

$$a_1 = (-1)^{(n-1)/2} \frac{(n+1)!}{2^n(\frac{n-1}{2})!(\frac{n+1}{2})!}$$

when n is odd. Thus we arrive at the following representation of Legendre's polynomial of degree n,

$$P_n(x) = \frac{1}{2^n} \sum_{k=0}^{[n/2]} (-1)^k \frac{(2n-2k)!}{k!(n-k)!(n-2k)!} x^{n-2k}, \tag{4.11}$$

where $[n/2]$ is the integral part of $n/2$, namely $n/2$ if n is even and $(n-1)/2$ if n is odd. The first six Legendre polynomials are (see Figure 4.1)

$$P_0(x) = 1, \qquad P_1(x) = x,$$

$$P_2(x) = \tfrac{1}{2}(3x^2 - 1), \qquad P_3(x) = \tfrac{1}{2}(5x^3 - 3x),$$

$$P_4(x) = \tfrac{1}{8}(35x^4 - 30x^2 + 3), \qquad P_5(x) = \tfrac{1}{8}(63x^5 - 70x^3 + 15x).$$

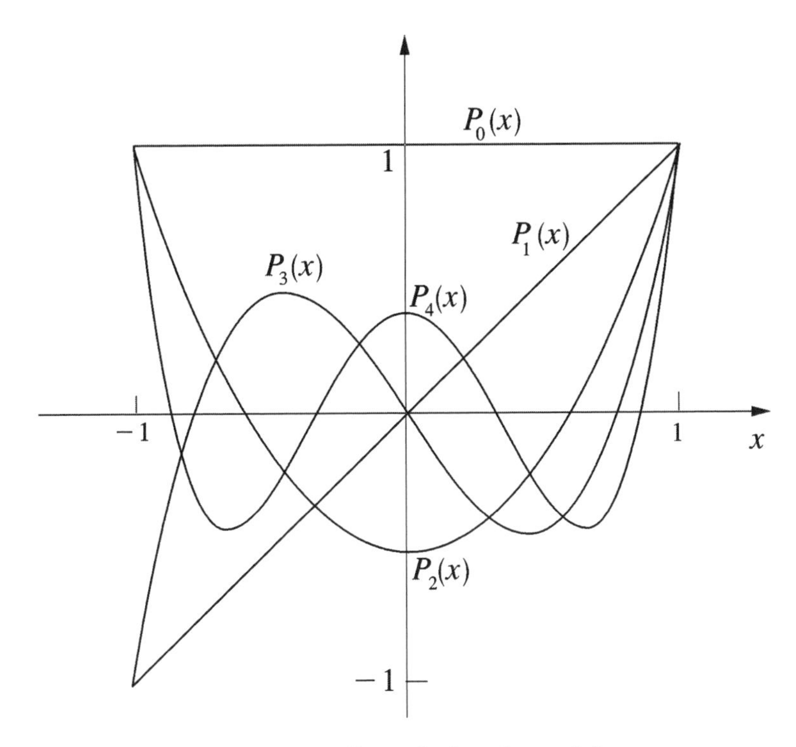

Figure 4.1 Legendre's polynomials.

The other solution of Legendre's equation, known as the *Legendre function* Q_n, is an infinite series which converges in the interval $(-1,1)$ and diverges outside it (see Exercise 4.3). The Legendre function of degree $n=0$ is given by

$$\begin{aligned} Q_0(x) &= x + \frac{1}{3}x^3 + \frac{1}{5}x^5 + \cdots \\ &= \frac{1}{2}\log\left(\frac{1+x}{1-x}\right). \end{aligned}$$

This becomes unbounded as x tends to ± 1 from within the interval $(-1,1)$. The same is true of $Q_1(x)$ (Exercises 4.3 and 4.4). The other Legendre functions Q_n can also be shown to be singular at the endpoints of the interval $(-1,1)$. The only eigenfunctions of Legendre's equation which are bounded at ± 1 are therefore the Legendre polynomials P_n. With $p(x) = 1 - x^2 = 0$ at $x = \pm 1$, the condition (4.2) is satisfied.

Thus the differential operator

$$-\frac{d}{dx}\left[(1-x^2)\frac{d}{dx}\right]$$

in Legendre's equation is self-adjoint. Its eigenvalues $\lambda_n = n(n+1)$ tend to ∞, and its eigenfunctions in $\mathcal{L}^2(-1,1)$, namely the Legendre polynomials P_n, are orthogonal and complete in $\mathcal{L}^2(-1,1)$, in accordance with Theorem 2.29. We verify the orthogonality of P_n directly in the next section.

EXERCISES

4.1 Verify that P_n satisfies Legendre's equation when $n=3$ and $n=4$.

4.2 Use the Gram–Schmidt method to construct an orthogonal set of polynomials out of the independent set

$$\{1, x, x^2, x^3, x^4, x^5 : -1 \leq x \leq 1\}.$$

Compare the result with the Legendre polynomials $P_0(x)$, $P_1(x)$, $P_2(x)$, $P_3(x)$, $P_4(x)$, and $P_5(x)$, and show that there is a linear relation between the two sets of polynomials.

4.3 Use the ratio test to prove that $(-1,1)$ is the interval of convergence for the power series which defines the Legendre function $Q_n(x)$. Show that the singularities of $Q_0(x)$ at $x = \pm 1$ are such that the product $(1-x^2)Q_0'(x)$ does not vanish at $x = \pm 1$, whereas $(1-x^2)Q_0(x) \to 0$ as $x \to \pm 1$, hence the boundary condition (4.2) is not satisfied when $u = Q_0$, $p(x) = 1 - x^2$, and v is a smooth function on $[-1,1]$.

4.4 Prove that

$$Q_1(x) = 1 - \frac{x}{2}\log\left(\frac{1+x}{1-x}\right).$$

4.5 Prove that, for all $n \in \mathbb{N}_0$,

$$\begin{aligned} P_n(-x) &= (-1)^n P_n(x), \\ P_{2n+1}(0) &= 0, \\ P_{2n}(0) &= (-1)^n \frac{(2n-1)(2n-3)\cdots(3)(1)}{(2n)(2n-2)\cdots(4)(2)}. \end{aligned}$$

4.6 Show that the substitution $x = \cos\theta$, $u(\cos\theta) = y(\theta)$ transforms Legendre's equation to

$$\sin\theta \frac{d^2y}{d\theta^2} + \cos\theta \frac{dy}{dx} + n(n+1)\sin\theta \; y = 0,$$

where $0 \leq \theta \leq \pi$. Note the appearance of the weight function $\sin\theta$ in this equation.

4.2 Properties of the Legendre Polynomials

For any positive integer n, we can write

$$\begin{aligned} (x^2-1)^n &= \sum_{k=0}^{n} (-1)^k \frac{n!}{k!(n-k)!} x^{2n-2k} \\ &= \sum_{k=0}^{[n/2]} (-1)^k \frac{n!}{k!(n-k)!} x^{2n-2k} \\ &\quad + \sum_{k=[n/2]+1}^{n} (-1)^k \frac{n!}{k!(n-k)!} x^{2n-2k}. \end{aligned} \tag{4.12}$$

Because

$$k = [n/2] + 1 = \begin{cases} n/2 + 1 & \text{if } n \text{ is even} \\ n/2 + 1/2 & \text{if } n \text{ is odd} \end{cases}$$

implies

$$2n - 2([n/2]+1) = \begin{cases} n-2 & \text{if } n \text{ is even} \\ n-1 & \text{if } n \text{ is odd,} \end{cases}$$

it follows that the powers on x in the second sum on the right-hand side of Equation (4.12) are all less than n. Taking the nth derivative of both sides of the equation, therefore, yields

$$\begin{aligned}\frac{d^n}{dx^n}(x^2-1)^n &= \frac{d^n}{dx^n}\sum_{k=0}^{[n/2]}(-1)^k \frac{n!}{k!(n-k)!}x^{2n-2k} \\ &= n!\sum_{k=0}^{[n/2]}(-1)^k \frac{(2n-2k)\cdots(n+1-2k)}{k!(n-k)!}x^{n-2k} \\ &= n!\sum_{k=0}^{[n/2]}(-1)^k \frac{(2n-2k)!}{k!(n-k)!(n-2k)!}x^{n-2k}.\end{aligned}$$

Comparing this with (4.11), we arrive at the Rodrigues formula for Legendre polynomials,

$$P_n(x) = \frac{1}{2^n n!}\frac{d^n}{dx^n}(x^2-1)^n, \tag{4.13}$$

which provides a convenient method for generating the first few polynomials:

$$\begin{aligned}P_0(x) &= 1, \\ P_1(x) &= \frac{1}{2}\frac{d}{dx}(x^2-1) = x, \\ P_2(x) &= \frac{1}{8}\frac{d^2}{dx^2}(x^2-1)^2 = \frac{1}{2}(3x^2-1), \ \ldots .\end{aligned}$$

It can also be used to derive some identities involving the Legendre polynomials and their derivatives, such as

$$P'_{n+1}(x) - P'_{n-1}(x) = (2n+1)P_n(x), \tag{4.14}$$

$$(n+1)P_{n+1}(x) + nP_{n-1}(x) = (2n+1)xP_n(x), \qquad n \in \mathbb{N}, \tag{4.15}$$

which are left as exercises.

We now use the Rodrigues formula to prove the orthogonality of the Legendre polynomials in $\mathcal{L}^2(-1,1)$. Suppose m and n are any two integers such that $0 \leq m < n$. Using integration by parts, we have

$$\begin{aligned}\int_{-1}^{1} P_n(x)x^m dx &= \frac{1}{2^n n!}\int_{-1}^{1} x^m \frac{d^n}{dx^n}(x^2-1)^n dx \\ &= \frac{1}{2^n n!}\left[x^m \frac{d^{n-1}}{dx^{n-1}}(x^2-1)^n\Big|_{-1}^{1} - m\int_{-1}^{1} x^{m-1}\frac{d^{n-1}}{dx^{n-1}}(x^2-1)^n dx\right] \\ &= \frac{-m}{2^n n!}\, x^{m-1}\frac{d^{n-2}}{dx^{n-2}}(x^2-1)^n\Big|_{-1}^{1} + \frac{m(m-1)}{2^n n!}\int_{-1}^{1} x^{m-2}\frac{d^{n-2}}{dx^{n-2}}(x^2-1)^n dx.\end{aligned}$$

Noting that all derivatives of $(x^2-1)^n$ of order less than n vanish at $x=\pm 1$, and repeating this process of integration by parts, we end up in the mth step with

$$\int_{-1}^{1} P_n(x)x^m dx = \frac{(-1)^m m!}{2^n n!} \frac{d^{n-m-1}}{dx^{n-m-1}}(x^2-1)^n \Big|_{-1}^{1} = 0,$$

Because $0 \leq n-m-1 < n$. Thus we see that P_n is orthogonal to x^m for all $m < n$, which implies that $P_n \perp P_m$ for all $m < n$. By symmetry we therefore conclude that

$$\langle P_n, P_m \rangle = 0 \qquad \text{for all } m \neq n.$$

To evaluate $\|P_n\|$, we use formula (4.13) once more to write

$$\|P_n\|^2 = \int_{-1}^{1} P_n^2(x)dx = \frac{1}{2^{2n}(n!)^2}\int_{-1}^{1} y^{(n)}(x)y^{(n)}(x)dx, \tag{4.16}$$

where $y(x) = (x^2-1)^n$, and then integrate by parts:

$$\begin{aligned}\int_{-1}^{1} y^{(n)}(x)y^{(n)}(x)dx &= -\int_{-1}^{1} y^{(n-1)}(x)y^{(n+1)}(x)dx \\ &= \cdots \\ &= (-1)^n \int_{-1}^{1} y(x)y^{(2n)}(x)dx \\ &= (-1)^n(2n)!\int_{-1}^{1} y(x)dx. \end{aligned} \tag{4.17}$$

$$\begin{aligned}(-1)^n\int_{-1}^{1}(x^2-1)^n dx &= \int_{-1}^{1}(1-x)^n(1+x)^n dx \\ &= \frac{n}{n+1}\int_{-1}^{1}(1-x)^{n-1}(1+x)^{n+1}dx \\ &= \cdots \\ &= \frac{n!}{(n+1)\cdots(2n)}\int_{-1}^{1}(1+x)^{2n}dx \\ &= \frac{(n!)^2}{(2n)!}\frac{(1+x)^{2n+1}}{2n+1}\Big|_{-1}^{1} \\ &= \frac{(n!)^2 2^{2n+1}}{(2n)!(2n+1)}. \end{aligned} \tag{4.18}$$

Combining Equations (4.16) through (4.18), we get

$$\|P_n\| = \sqrt{\frac{2}{2n+1}}, \qquad n \in \mathbb{N}_0,$$

and hence the sequence of polynomials

$$\frac{1}{\sqrt{2}}P_0(x),\ \sqrt{\frac{3}{2}}P_1(x),\ \sqrt{\frac{5}{2}}P_3(x),\ \ldots,\ \sqrt{\frac{2n+1}{2}}P_n(x),\ \ldots$$

is a complete orthonormal system in $\mathcal{L}^2(-1,1)$. Note how $\|P_n\|$, unlike $\|\sin nx\|$ and $\|\cos nx\|$, actually depends on n and tends to 0 as $n \to \infty$.

Example 4.1

Since the function

$$f(x) = \begin{cases} 0, & -1 < x < 0 \\ 1, & 0 < x < 1 \end{cases}$$

is square integrable on $(-1,1)$, it can be represented in $\mathcal{L}^2(-1,1)$ by the Fourier–Legendre series

$$f(x) = \sum_{n=0}^{\infty} c_n P_n(x),$$

in which the Fourier–Legendre coefficients c_n are determined, according to the expansion formula (4.3), by

$$c_n = \frac{\langle f, P_n \rangle}{\|P_n\|^2} = \frac{2n+1}{2}\int_0^1 P_n(x)dx.$$

The series $\sum c_n P_n$ is also referred to as a Legendre series and c_n its Legendre coefficients. Thus

$$f(x) = \frac{1}{2}P_0(x) + \frac{3}{4}P_1(x) - \frac{7}{16}P_3(x) + \cdots. \tag{4.19}$$

Observe that, in as much as

$$f(x) - \frac{1}{2}P_0(x) = f(x) - \frac{1}{2} = \begin{cases} -1/2, & -1 < x < 0 \\ 1/2, & 0 < x < 1 \end{cases}$$

is an odd function, the Legendre series representation of f does not contain any even Legendre polynomials except P_0. Furthermore, because $P_n(0) = 0$ for every odd n, the right-hand side of (4.19) at $x = 0$ equals

$$\frac{1}{2}P_0(0) = \frac{1}{2} = \frac{1}{2}[f(0^+) + f(0^-)],$$

in agreement with the generalized version of Theorem 3.9 referred to at the beginning of this chapter.

The Legendre polynomials are generated by the function $(1-2xt+t^2)^{-1/2}$ in the sense that

$$\frac{1}{\sqrt{1-2xt+t^2}} = \sum_{n=0}^{\infty} P_n(x)t^n, \qquad |t| < 1, \quad |x| \leq 1. \tag{4.20}$$

To prove this identity we note that, for every fixed x in $[-1,1]$, the left-hand side of (4.20) is an analytic function of t in the interval $(-1,1)$, which we denote

$$f(t,x) = \sum_{n=0}^{\infty} a_n(x)t^n, \qquad |t| < 1. \tag{4.21}$$

This function satisfies the differential equation

$$(1-2xt+t^2)\frac{\partial f}{\partial t} = (x-t)f. \tag{4.22}$$

Substituting the series representation (4.21) into (4.22) leads to the recursion relation

$$(n+1)a_{n+1}(x) + na_{n-1}(x) = (2n+1)xa_n(x), \qquad n \in \mathbb{N}_0,$$

which is the same as the relation (4.15) between P_{n+1}, P_{n-1}, and P_n. The first two coefficients in the Taylor series expansion of $(1-2xt+t^2)^{1/2}$ are

$$a_0(x) = 1 = P_0(x), \quad a_1(x) = x = P_1(x),$$

therefore it follows that $a_n(x) \equiv P_n(x)$ for all $n \in \mathbb{N}_0$.

Setting $x = 1$ and $x = -1$ in (4.20), we obtain

$$\sum_{n=0}^{\infty} P_n(1)t^n = \frac{1}{\sqrt{1-2t+t^2}} = \frac{1}{1-t} = \sum_{n=0}^{\infty} t^n,$$

$$\sum_{n=0}^{\infty} P_n(-1)t^n = \frac{1}{\sqrt{1+2t+t^2}} = \frac{1}{1+t} = \sum_{n=0}^{\infty}(-1)^n t^n.$$

Hence

$$P_n(1) = 1, \; P_n(-1) = (-1)^n \quad \text{for all } n \in \mathbb{N}_0.$$

EXERCISES

4.7 Use the Rodrigues formula to prove the identity (4.14), then conclude that

$$\int_{\pm 1}^{x} P_n(t)dt = \frac{1}{2n+1}[P_{n+1}(x) - P_{n-1}(x)],$$

$$\int_{-1}^{1} P_n(t)dt = 0 \quad \text{for all } n \in \mathbb{N}.$$

4.8 Prove the identity (4.15).

4.9 Show in detail how Equation (4.22) follows from Equation (4.21), and how the two equations imply the recursion relation for a_n.

4.10 Show that if $0 \le r < 1$, then

$$\frac{1}{|1 - re^{i\theta}|} = \sum_{n=0}^{\infty} P_n(\cos\theta) r^n.$$

Hence deduce that, if $\mathbf{x}$ and $\mathbf{y}$ are nonzero vectors in $\mathbb{R}^3$ such that $r = \|\mathbf{x}\| / \|\mathbf{y}\| < 1$ and θ is the angle between them, then

$$\frac{1}{\|\mathbf{y} - \mathbf{x}\|} = \frac{1}{\|\mathbf{y}\|} \sum_{n=0}^{\infty} P_n(\cos\theta) r^n.$$

In particular, if $\|\mathbf{y}\| = 1$, then

$$\frac{1}{\|\mathbf{y} - \mathbf{x}\|} = \sum_{n=0}^{\infty} P_n(\cos\theta) \|\mathbf{x}\|^n.$$

Thus, if two masses are located at $\mathbf{x}$ and $\mathbf{y}$, then the gravitational potential between them may be represented (up to a multiplicative constant) by a power series in $\|\mathbf{x}\|$ with coefficients $P_n(\cos\theta)$.

4.11 Expand each of the following functions in Legendre polynomials.

(a) $f(x) = 1 - x^3$, $-1 \le x \le 1$.

(b) $f(x) = |x|$, $-1 \le x \le 1$.

Compute the first six terms ($n = 0$ up to $n = 5$) in (b).

4.12 Use Equation (4.15) to prove that

$$n\|P_n\|^2 = (2n-1)\langle xP_{n-1}, P_n\rangle, \quad n\|P_{n-1}\|^2 = (2n+1)\langle xP_n, P_{n-1}\rangle,$$

and hence conclude that $\|P_n\|^2 = 2/(2n+1)$.

4.13 Expand the function

$$f(x) = \begin{cases} -1, & -1 < x < 0 \\ 1, & 0 < x < 1 \end{cases}$$

in a Legendre series. Use the formulas in Exercise 4.7 to evaluate the coefficients. Compute the value of the series at $x = 0$ and compare it with the average value of $f(0^+)$ and $f(0^-)$.

4.14 Give two Legendre series expansions of the function $f(x) = x$ on $[0, 1]$, one in even degree polynomials, and the other in odd degree. Is the expansion valid at $x = 0$? Repeat the procedure if the function is $f(x) = 1$ on $[0, 1]$.

4.3 Hermite and Laguerre Polynomials

4.3.1 Hermite Polynomials

Legendre's polynomials were defined as a set of solutions of a certain differential equation (Legendre's equation). The same procedure may be followed to define the Hermite polynomials. But, instead, we start by defining these polynomials, then deduce the equation (Hermite's equation) that they satisfy, although this is not the natural order in which the theory developed; for both equations arise naturally in mathematical physics, as we show later in this chapter.

For each $n \in \mathbb{N}_0$, the function $H_n : \mathbb{R} \to \mathbb{R}$ is defined by

$$H_n(x) = (-1)^n e^{x^2} \frac{d^n}{dx^n} e^{-x^2}. \tag{4.23}$$

This yields a sequence of polynomials

$$\begin{aligned} H_0(x) &= 1, & H_1(x) &= 2x, \\ H_2(x) &= 4x^2 - 2, & H_3(x) &= 8x^3 - 12x, \\ H_4(x) &= 16x^4 - 48x^2 + 12, & H_5(x) &= 32x^5 - 160x^3 + 120x, \\ &\vdots & &\vdots \end{aligned}$$

known as Hermite polynomials, named after the French mathematician Charles Hermite (1822–1901). Using the formula (4.23), we arrive at the following properties for the set $\{H_n : n \in \mathbb{N}_0\}$.

1. H_n is a polynomial of degree n.

Proof

$$\begin{aligned}
\frac{d}{dx}e^{-x^2} &= -2xe^{-x^2} \\
\frac{d^2}{dx^2}e^{-x^2} &= (-2x)^2e^{-x^2} - 2e^{-x^2} \\
\frac{d^3}{dx^3}e^{-x^2} &= (-2x)^3e^{-x^2} + 12xe^{-x^2} \\
&\vdots
\end{aligned}$$

By induction, we obtain

$$\frac{d^n}{dx^n}e^{-x^2} = (-2x)^n e^{-x^2} + p(x)e^{-x^2},$$

where p is a polynomial of degree less than n. Therefore

$$\begin{aligned}
H_n(x) &= (-1)^n e^{x^2}[(-2x)^n + p(x)]e^{-x^2} \\
&= (2x)^n + (-1)^n p(x). \qquad (4.24)
\end{aligned}$$

□

2. The set $\{H_n : n \in \mathbb{N}_0\}$ is orthogonal in $\mathcal{L}^2_{e^{-x^2}}(\mathbb{R})$.

Proof

Assuming $m < n$,

$$\begin{aligned}
\langle H_m, H_n\rangle_{e^{-x^2}} &= \int_{-\infty}^{\infty} H_m(x)H_n(x)e^{-x^2}dx \\
&= (-1)^n \int_{-\infty}^{\infty} H_m(x)\frac{d^n}{dx^n}e^{-x^2}dx.
\end{aligned}$$

Integrating by parts n times, and noting that, for any polynomial p, $p(x)e^{-x^2} \to 0$ as $|x| \to \infty$, we obtain

$$\langle H_m, H_n\rangle_{e^{-x^2}} = (-1)^{2n} \int_{-\infty}^{\infty} \left[\frac{d^n}{dx^n}H_m(x)\right] e^{-x^2}dx = 0.$$

Consequently, H_m is orthogonal to H_n for all $m < n$ and hence, by the symmetry of the inner product, for all $m \neq n$. □

3. $\|H_n\|^2 = 2^n n! \sqrt{\pi}$.

Proof

$$\begin{aligned}
\|H_n\|^2 &= \int_{-\infty}^{\infty} H_n^2(x) e^{-x^2} dx \\
&= (-1)^n \int_{-\infty}^{\infty} H_n(x) \frac{d^n}{dx^n} e^{-x^2} dx \\
&= \int_{-\infty}^{\infty} \left[\frac{d^n}{dx^n} H_n(x) \right] e^{-x^2} dx \\
&= \int_{-\infty}^{\infty} 2^n n! e^{-x^2} dx,
\end{aligned}$$

where the last equality follows from Equation (4.24). But

$$\int_{-\infty}^{\infty} e^{-x^2} dx = \sqrt{\pi}$$

(see Exercise 4.15), hence we arrive at the desired equality. □

4. For every $x \in \mathbb{R}$,

$$e^{2xt-t^2} = \sum_{n=0}^{\infty} \frac{1}{n!} H_n(x) t^n. \tag{4.25}$$

In other words, e^{2xt-t^2} is a generating function for the Hermite polynomials.

Proof

For every $x \in \mathbb{R}$ the function $f(t,x) = e^{2xt-t^2}$ is analytic in t over $\mathbb{R}$, and is therefore represented by the power series

$$f(t,x) = \sum_{n=0}^{\infty} \frac{1}{n!} \left. \frac{\partial^n f}{\partial t^n} \right|_{t=0} t^n.$$

Because

$$\begin{aligned}
\left. \frac{\partial^n f}{\partial t^n} \right|_{t=0} &= \left. \frac{\partial^n}{\partial t^n} e^{x^2-(x-t)^2} \right|_{t=0} \\
&= e^{x^2} \left. \frac{\partial^n}{\partial t^n} e^{-(x-t)^2} \right|_{t=0} \\
&= (-1)^n e^{x^2} \left. \frac{d^n}{dy^n} e^{-y^2} \right|_{y=x} \\
&= (-1)^n e^{x^2} \frac{d^n}{dx^n} e^{-x^2},
\end{aligned}$$

where $y = x - t$, it follows that

$$\left.\frac{\partial^n f}{\partial t^n}\right|_{t=0} = (-1)^n e^{x^2} \frac{d^n}{dx^n} e^{-x^2} = H_n(x).$$

□

Theorem 4.2

H_n satisfies the second-order differential equation

$$u'' - 2xu' + 2nu = 0, \quad x \in \mathbb{R}. \tag{4.26}$$

Proof

Differentiating the identity (4.25) with respect to x, we obtain

$$2te^{2xt-t^2} = \sum_{n=1}^{\infty} \frac{1}{n!} H_n'(x) t^n, \tag{4.27}$$

where the left-hand side can also be written as

$$\begin{aligned} 2te^{2xt-t^2} &= 2\sum_{n=0}^{\infty} \frac{1}{n!} H_n(x) t^{n+1} \\ &= 2\sum_{n=0}^{\infty} \frac{n+1}{(n+1)!} H_n(x) t^{n+1} \\ &= 2\sum_{n=1}^{\infty} \frac{n}{n!} H_{n-1}(x) t^n. \end{aligned} \tag{4.28}$$

By comparing Equations (4.27) and (4.28) we arrive at the relation

$$H_n'(x) = 2nH_{n-1}(x), \quad n \in \mathbb{N}. \tag{4.29}$$

Differentiating (4.25) with respect to t gives

$$2(x-t)e^{2xt-t^2} = \sum_{n=1}^{\infty} \frac{1}{(n-1)!} H_n(x) t^{n-1} = \sum_{n=0}^{\infty} \frac{1}{n!} H_{n+1}(x) t^n,$$

from which

$$\begin{aligned} 2x\sum_{n=0}^{\infty} \frac{1}{n!} H_n(x) t^n &= 2\sum_{n=0}^{\infty} \frac{1}{n!} H_n(x) t^{n+1} + \sum_{n=0}^{\infty} \frac{1}{n!} H_{n+1}(x) t^n \\ &= 2\sum_{n=0}^{\infty} \frac{n+1}{(n+1)!} H_n(x) t^{n+1} + \sum_{n=0}^{\infty} \frac{1}{n!} H_{n+1}(x) t^n \\ &= 2\sum_{n=1}^{\infty} \frac{n}{n!} H_{n-1}(x) t^n + \sum_{n=0}^{\infty} \frac{1}{n!} H_{n+1}(x) t^n. \end{aligned}$$

Comparing the two sides of this equation, we see that

$$\begin{aligned} 2xH_0(x) &= H_1(x) \\ 2xH_n(x) &= 2nH_{n-1}(x) + H_{n+1}(x), \quad n \in \mathbb{N}. \end{aligned} \tag{4.30}$$

Combining Equations (4.29) and (4.30) gives

$$2xH_n(x) = H_n'(x) + H_{n+1}(x),$$

and differentiating this last equation, we end up with

$$\begin{aligned} H_n''(x) &= 2xH_n'(x) + 2H_n(x) - H_{n+1}'(x) \\ &= 2xH_n'(x) + 2H_n(x) - (2n+2)H_n(x) \\ &= 2xH_n'(x) - 2nH_n(x). \end{aligned}$$

□

Equation (4.26) is called Hermite's equation, and we have shown that one solution of this equation is the Hermite polynomial H_n. As with Legendre's equation, the other solution turns out to be an analytic function which can be represented by a power series in x (see Exercise (4.21)). Multiplication by e^{-x^2} converts Hermite's equation to the standard Sturm–Liouville form

$$(e^{-x^2}u')' + 2ne^{-x^2}u = 0, \quad x \in \mathbb{R}, \tag{4.31}$$

in which $p(x) = e^{-x^2}$, $\lambda = 2n$, $\rho(x) = e^{-x^2}$, and the differential operator

$$\frac{d}{dx}\left(e^{-x^2}\frac{d}{dx}\right)$$

is formally self-adjoint. Because $H_n(x)$, but not the infinite series solution (see Exercise 4.22), lies in $\mathcal{L}^2_{e^{-x^2}}(\mathbb{R})$, we therefore conclude that the sequence of Hermite polynomials $(H_n : n \in \mathbb{N}_0)$ forms a complete orthogonal system in $\mathcal{L}^2_{e^{-x^2}}(\mathbb{R})$.

4.3.2 Laguerre Polynomials

The differential equation

$$xu'' + (1-x)u' + nu = 0, \quad 0 < x < \infty, \quad n \in \mathbb{N}_0, \tag{4.32}$$

is known as Laguerre's equation, after the French mathematician Edmond Laguerre (1834–1886). Its standard form

$$(xe^{-x}u')' + ne^{-x}u = 0 \tag{4.33}$$

is obtained by multiplication by e^{-x}, the weight function in this case. Note that $p(x) = xe^{-x}$ vanishes at $x = 0$, so no boundary condition is needed at this point. The nonsingular solutions of Laguerre's equation are given, up to an arbitrary multiplicative constant, by the Laguerre polynomials

$$\begin{aligned} L_0(x) &= 1, \\ L_1(x) &= 1 - x, \\ L_2(x) &= 1 - 2x + \frac{1}{2}x^2, \\ &\vdots \\ L_n(x) &= \frac{e^x}{n!}\frac{d^n}{dx^n}(x^n e^{-x}), \end{aligned} \tag{4.34}$$

where L_n has degree n. Because L_n are eigenfunctions of the SL operator in Equation (4.33), we would expect to have $\langle L_m, L_n \rangle_{e^{-x}} = 0$ for all $m \neq n$, and this follows immediately from the observation that, for all $m < n$,

$$\begin{aligned} \langle x^m, L_n \rangle_{e^{-x}} &= \int_0^\infty x^m L_n(x) e^{-x} dx \\ &= \frac{1}{n!}\int_0^\infty x^m \frac{d^n}{dx^n}(x^n e^{-x}) dx \\ &= (-1)^m \frac{m!}{n!}\int_0^\infty \frac{d^{n-m}}{dx^{n-m}}(x^n e^{-x}) dx = 0. \end{aligned}$$

Furthermore,

$$\langle x^n, L_n \rangle_{e^{-x}} = (-1)^n \int_0^\infty x^n e^{-x} dx = (-1)^n n!,$$

and because the coefficient of x^n in L_n is $(-1)^n/n!$, it follows that

$$\|L_n\|^2 = \langle L_n, L_n \rangle = \frac{(-1)^n}{n!}\langle x^n, L_n \rangle = 1$$

for all n, and the Laguerre polynomials are in fact orthonormal (and complete) in $\mathcal{L}^2_{e^{-x}}(0, \infty)$.

EXERCISES

4.15 Show that $\int_{-\infty}^{\infty} e^{-x^2} dx = \sqrt{\pi}$ by using the equality

$$\left(\int_0^\infty e^{-x^2} dx\right)^2 = \int_0^\infty \int_0^\infty e^{-(x^2+y^2)} dx dy.$$

Hint: Evaluate the double integral by transforming to polar coordinates.

4.16 Show that the powers of x in the polynomial $H_n(x)$ are even if n is even, and odd if n is odd.

4.17 Prove that, for all $n \in \mathbb{N}_0$,

$$H_{2n}(0) = (-1)^n \frac{(2n)!}{n!}, \quad H_{2n+1}(0) = 0.$$

4.18 Use induction to prove

$$H_n(x) = n! \sum_{k=0}^{[n/2]} (-1)^k \frac{(2x)^{n-2k}}{k!(n-2k)!}.$$

4.19 Expand the function $f(x) = x^m$, $m \in \mathbb{N}_0$, in Hermite polynomials.

4.20 Expand the function

$$f(x) = \begin{cases} 0, & x < 0 \\ 1, & x > 0 \end{cases}$$

in terms of Hermite polynomials. Compute the first five terms.

4.21 The differential equation

$$u'' - 2xu' + 2\lambda u = 0,$$

in which λ is not necessarily an integer, is a generalized version of Equation (4.26), and is also called Hermite's equation.

(a) Assuming that $u(x) = \sum_{k=0}^{\infty} c_k x^{k+r}$ and substituting into the equation, show that

$$c_{k+2} = \frac{2(k+r-\lambda)}{(k+r+2)(k+r+1)} c_k, \quad k \in \mathbb{N}_0.$$

(b) If $r = 0$, the solution is

$$u_0(x) = 1 - \frac{2\lambda}{2!}x^2 + \frac{2^2\lambda(\lambda-2)}{4!}x^4 - \cdots.$$

If $r = 1$, we get

$$u_1(x) = x - \frac{2(\lambda-1)}{3!}x^3 + \frac{2^2(\lambda-1)(\lambda-3)}{5!}x^5 - \cdots.$$

(c) Conclude that the general solution of Hermite's equation for any real λ is $u(x) = c_0 u_0(x) + c_1 u_1(x)$, where c_0 and c_1 are constants. What happens when λ is a nonnegative integer?

4.22 With u_0 and u_1 as in Exercise 4.21, prove that each of the functions $e^{-x^2}u_0(x)$ and $e^{-x^2}u_1(x)$ tends to a constant as $|x| \to \infty$, and that this constant is 0 if, and only if, the solution is H_n. Conclude from this that the only solutions of Hermite's equation in $\mathcal{L}^2_{e^{-x^2}}(\mathbb{R})$ are the Hermite polynomials.

4.23 Prove that

$$L_n(x) = \sum_{k=0}^{n} (-1)^k \frac{1}{k!} \binom{n}{k} x^k,$$

where

$$\binom{n}{k} = \frac{n!}{k!(n-k)!}$$

are the binomial coefficients.

4.24 Express $x^3 - x$ as a linear combination of Laguerre polynomials.

4.25 Expand $f(x) = x^m$, where m is a positive integer, in terms of Laguerre polynomials.

4.26 Expand $f(x) = e^{x/2}$ in a Laguerre series on $[0, \infty)$.

4.27 Verify that L_n satisfies Equation (4.32).

4.28 Determine the general solution of Laguerre's equation when $n = 0$.

4.4 Physical Applications

The orthogonal polynomials discussed in this chapter have historically been intimately tied up with the study of potential fields and, more recently, quantum mechanics. Here we look at two typical examples where the Legendre and Hermite polynomials crop up. The first is in the description of an electrostatic field generated by a spherical capacitor, as a result of solving Laplace's equation in spherical coordinates. The second is in representing the wave function for a harmonic oscillator.

4.4.1 Laplace's Equation

In $\mathbb{R}^3$ the second-order partial differential equation

$$\frac{\partial^2 u}{\partial x^2} + \frac{\partial^2 u}{\partial y^2} + \frac{\partial^2 u}{\partial z^2} = 0, \tag{4.35}$$

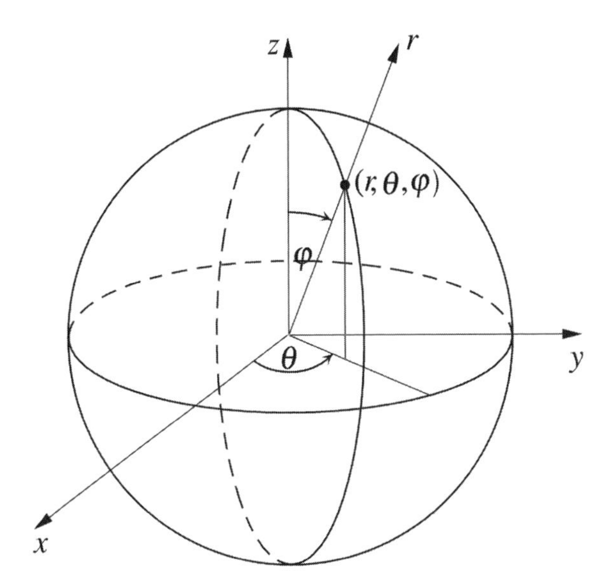

Figure 4.2 Spherical coordinates.

is called Laplace's equation, after the French mathematician Pierre de Laplace (1749–1827). It provides a mathematical model for a number of significant physical phenomena, such as the static distribution of an electric, gravitational, or temperature field in free space, as pointed out in Section 3.3. In this context the scalar function $u = u(x, y, z)$ represents the electric potential, the gravitational potential, or the temperature at the point (x, y, z). Under the transformation from Cartesian coordinates (x, y, z) to spherical coordinates (r, θ, φ), defined by (see Figure 4.2)

$$\begin{aligned} x &= r\cos\theta\sin\varphi \\ y &= r\sin\theta\sin\varphi \\ z &= r\cos\varphi, \end{aligned}$$

where $r \geq 0$, $-\pi < \theta \leq \pi$, $0 \leq \varphi \leq \pi$, Equation (4.35) takes the form

$$\frac{\partial}{\partial r}\left(r^2\frac{\partial u}{\partial r}\right) + \frac{1}{\sin^2\varphi}\frac{\partial^2 u}{\partial\theta^2} + \frac{1}{\sin\varphi}\frac{\partial}{\partial\varphi}\left(\sin\varphi\frac{\partial u}{\partial\varphi}\right) = 0. \tag{4.36}$$

Assuming that the function u is symmetric with respect to the z-axis, it will be independent of θ and Equation (4.36) becomes

$$\frac{\partial}{\partial r}\left(r^2\frac{\partial u}{\partial r}\right) + \frac{1}{\sin\varphi}\frac{\partial}{\partial\varphi}\left(\sin\varphi\frac{\partial u}{\partial\varphi}\right) = 0. \tag{4.37}$$

We use the method of separation of variables to solve Equation (4.37), so we assume that $u(r, \varphi)$ is a product of a function of r and a function of φ,

$$u(r, \varphi) = v(r)w(\varphi),$$

then substitute into (4.37). After division by vw this gives

$$\frac{1}{v}\frac{d}{dr}\left(r^2\frac{dv}{dr}\right) = -\frac{1}{w\sin\varphi}\frac{d}{d\varphi}\left(\sin\varphi\frac{dw}{d\varphi}\right), \tag{4.38}$$

where the left-hand side depends only on r, and the right-hand side depends only on φ. Assuming Equation (4.38) holds over a domain Ω (open, connected set) of the $r\varphi$-plane, each side must be a constant, which we denote λ. Thus we obtain the pair of equations

$$r^2v'' + 2rv' - \lambda v = 0, \tag{4.39}$$

$$\frac{1}{\sin\varphi}(\sin\varphi w')' + \lambda w = 0. \tag{4.40}$$

Setting $\xi = \cos\varphi$ in Equation (4.40), and noting that

$$\frac{d}{d\varphi} = \frac{d\xi}{d\varphi}\frac{d}{d\xi} = -\sin\varphi\frac{d}{d\xi},$$

$$\frac{1}{\sin\varphi}\frac{d}{d\varphi}\left(\sin\varphi\frac{dw}{d\varphi}\right) = \frac{d}{d\xi}\left[(1-\xi^2)\frac{dw}{d\xi}\right],$$

leads to Legendre's equation

$$\frac{d}{d\xi}\left[(1-\xi^2)\frac{dw}{d\xi}\right] + \lambda w = 0.$$

If $\lambda = n(n+1)$, $n \in \mathbb{N}_0$, then the solutions of Equation (4.40) are given by the Legendre polynomials

$$w_n(\varphi) = P_n(\xi) = P_n(\cos\varphi).$$

Equation (4.39), on the other hand, is of the Cauchy–Euler type and the substitution $v(r) = r^\alpha$ yields

$$[\alpha(\alpha-1) + 2\alpha - n(n+1)]r^\alpha = 0,$$

which implies $\alpha = n$ or $\alpha = -n-1$, hence

$$v_n(r) = a_nr^n + b_nr^{-n-1}$$

for some arbitrary constants a_n and b_n. Thus the sequence

$$u_n(r, \varphi) = v_n(r)w_n(\varphi) = (a_nr^n + b_nr^{-n-1})P_n(\cos\varphi)$$

satisfies Laplace's equation (4.37) for each $n \in \mathbb{N}_0$, and the complete solution of the equation is therefore formally given by

$$\begin{aligned} u(r,\varphi) &= \sum_{n=0}^{\infty} u_n(r,\varphi) \\ &= \sum_{n=0}^{\infty} (a_n r^n + b_n r^{-n-1}) P_n(\cos\varphi). \end{aligned} \tag{4.41}$$

Two observations are worth noting in connection with the representation (4.41) of the solution to Laplace's equation. The first is that the Legendre functions $Q_n(\cos\varphi)$ were not taken into account, because these functions become unbounded along the z-axis (where $\cos\varphi = \pm 1$). For the same reason the terms $b_n r^{-n-1}$ are dropped (by setting $b_n = 0$) if the domain Ω includes the origin $r = 0$. The second observation, already mentioned in Remark 3.16, is that the generality lost in the assumption that $u(r,\varphi)$ is a product of a function of r and a function of φ is restored by taking a linear combination of these solutions as we do in (4.41), P_n being complete in $\mathcal{L}^2(-1,1)$. Of course the constants a_n and b_n are determined by additional (boundary) conditions.

Suppose, for example, that $u(r,\varphi)$ satisfies Laplace's equation inside the ball $0 \le r < R$ and that $u(R,\varphi) = f(\varphi)$ on the spherical surface $r = R$, where f is a given function in $\mathcal{L}^2(0,\pi)$. Using (4.41) with $b_n = 0$ for all n,

$$u(r,\varphi) = \sum_{n=0}^{\infty} a_n r^n P_n(\cos\varphi). \tag{4.42}$$

Applying the boundary condition at $r = R$, we obtain

$$u(R,\varphi) = \sum_{n=0}^{\infty} a_n R^n P_n(\cos\varphi) = f(\varphi),$$

from which we conclude that $a_n R^n$ are the Fourier–Legendre coefficients of $f(\varphi)$ when expanded in terms of $P_n(\cos\varphi)$. Thus

$$\begin{aligned} a_n R^n &= \frac{\langle f, P_n \rangle}{\|P_n\|^2} \\ &= \frac{2n+1}{2} \int_{-1}^{1} f(\varphi(\xi)) P_n(\xi) d\xi, \end{aligned}$$

and therefore

$$a_n = \frac{2n+1}{2R^n} \int_0^{\pi} f(\varphi) P_n(\cos\varphi) \sin\varphi \, d\varphi, \quad n \in \mathbb{N}_0. \tag{4.43}$$

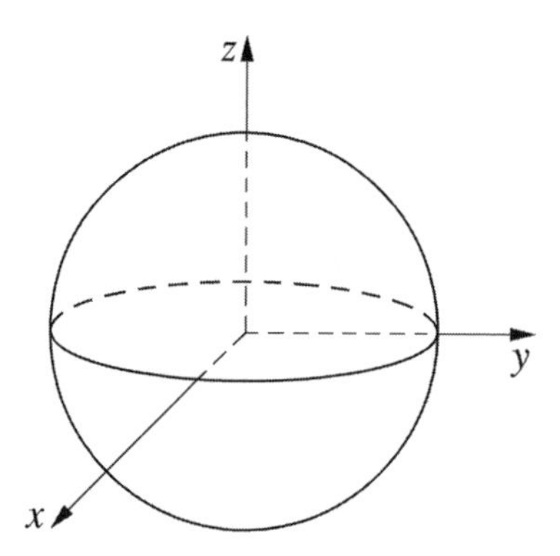

Figure 4.3 Spherical capacitor.

Outside the ball, in the region $r > R$, the nonnegative powers of r become unbounded and their coefficients must therefore vanish. Hence the solution in this case is given by

$$u(r, \varphi) = \sum_{n=0}^{\infty} b_n r^{-n-1} P_n(\cos\varphi),$$

where

$$b_n = \frac{2n+1}{2} R^{n+1} \int_0^{\pi} f(\varphi) P_n(\cos\varphi) \sin\varphi \, d\varphi.$$

Example 4.3

If opposite electric charges are placed on two hemispherical conducting sheets, which are insulated from each other, an electric field is generated between them, both inside and outside the spherical surface. The apparatus is called an *electric capacitor* (see Figure 4.3). Suppose the radius of the spherical capacitor is 1. If the upper hemisphere has potential 1, and the lower hemisphere has potential -1, determine the potential function inside the sphere.

Solution

Let u denote the potential function of the electric field at any point in space outside the conducting surfaces. The electric potential (and hence the electric charge) on each hemisphere is distributed symmetrically about the z-axis, therefore u depends only on r and φ. Inside the sphere it has the form (4.42). On the spherical surface,

$$u(1,\varphi)=\begin{cases}1, & 0\leq\varphi<\pi/2\\ -1, & \pi/2<\varphi\leq\pi.\end{cases}$$

Applying the formula (4.43), with $R=1$ and $f(\varphi)=u(1,\varphi)$,

$$\begin{aligned}a_n&=\frac{2n+1}{2}\left[\int_0^{\pi/2}P_n(\cos\varphi)\sin\varphi\,d\varphi+\int_{\pi/2}^{\pi}(-1)P_n(\cos\varphi)\sin\varphi\,d\varphi\right]\\&=\frac{2n+1}{2}\int_0^1[P_n(\xi)-P_n(-\xi)]d\xi.\end{aligned}$$

Therefore $a_n=0$ for even values of n, and

$$a_n=\frac{2n+1}{2}\int_0^1 2P_n(\xi)d\xi=(2n+1)\int_0^1 P_n(\xi)d\xi$$

when n is odd. Thus

$$\begin{aligned}a_1&=3\int_0^1 P_1(\xi)d\xi=\frac{3}{2}\\a_3&=7\int_0^1 P_3(\xi)d\xi=-\frac{7}{8}\\a_5&=11\int_0^1 P_5(\xi)d\xi=\frac{11}{16}\\&\vdots\end{aligned}$$

Based on the result of Exercise 4.7, a general formula for a_n, when n is odd, is given by

$$\begin{aligned}a_n&=(2n+1)\int_0^1 P_n(\xi)d\xi\\&=P_{n-1}(0)-P_{n+1}(0).\end{aligned}$$

The potential inside the capacitor is therefore

$$u(r,\varphi)=\frac{3}{2}rP_1(\cos\varphi)-\frac{7}{8}r^3P_3(\cos\varphi)+\frac{11}{16}r^5P_5(\cos\varphi)+\cdots.$$

4.4.2 Harmonic Oscillator

In quantum mechanics the state of a particle which is constrained to move on a straight line (the x-axis) is described by a wave function $\Psi(x,t)$. If the motion is harmonic, the dependence on time t is given by

$$\Psi(x,t)=\psi(x)e^{-iEt/\hbar},$$

where E is the total energy of the particle and $\hbar$ is a universal constant known as Plank's constant. The location of the particle is then determined by the function $\psi(x)$ in the sense that $|\psi(x)|^2 / \|\psi\|^2$ is a measure of its probability density in $\mathbb{R}$. If the particle is located in a potential field $V(x)$, the function ψ satisfies the time-independent Schrödinger equation

$$\psi'' + \frac{2m}{\hbar^2}[E - V(x)]\psi = 0.$$

By defining $\lambda = 2mE/\hbar^2$ and $r(x) = -2mV(x)/\hbar^2$, Schrödinger's equation takes the form of a standard Sturm–Liouville eigenvalue equation

$$\psi'' + [\lambda + r(x)]\psi = 0. \tag{4.44}$$

On the interval $-\infty < x < \infty$, the condition that $\psi \to 0$ as $x \to \pm\infty$ defines a singular SL problem.

When $V(x)$ is proportional to x^2, the resulting system is called a *harmonic oscillator*. For the sake of simplicity we take $V(x) = \hbar^2 x^2/2m$, so that Equation (4.44) becomes

$$\psi'' + (\lambda - x^2)\psi = 0. \tag{4.45}$$

If we set $u(x) = e^{x^2/2}\psi(x)$, it is a simple matter to show that

$$u'' - 2xu' + (\lambda - 1)u = 0,$$

which is Hermite's equation when $\lambda = 2n + 1$. Its solution is therefore H_n. Consequently, the eigenvalues of Equation (4.45) are

$$\lambda_n = 2n + 1, \quad n \in \mathbb{N}_0,$$

which determine the (admissible) energy levels $E_n = (2n + 1)\hbar^2/2m$ of the particle in the harmonic oscillator, and the corresponding eigenfunctions are given by

$$\psi_n(x) = e^{-x^2/2}H_n(x), \quad n \in \mathbb{N}_0,$$

which represent the wave functions of the particle in the admissible energy levels. The functions $e^{-x^2/2}H_n(x)$, which belong to $\mathcal{L}^2(\mathbb{R})$, are called Hermite functions. The probability that the particle is located in the interval (a, b) is given by

$$\frac{1}{\|H_n\|_{e^{-x^2}}^2}\int_a^b H_n^2(x)e^{-x^2}dx = \frac{\int_a^b H_n^2(x)e^{-x^2}dx}{\int_{-\infty}^{\infty} H_n^2(x)e^{-x^2}dx}.$$

EXERCISES

4.29 In Example 4.3, determine the surface where $u(r, \varphi) = 0$.

4.30 In Example 4.3, determine the potential function outside the sphere $r = 1$.

4.31 Find the solution $u(r, \varphi)$ of Laplace's equation inside a sphere of radius R if

$$u(R, \varphi) = \begin{cases} 10, & 0 \leq \varphi < \pi/2 \\ 0, & \pi/2 < \varphi \leq \pi. \end{cases}$$

4.32 Suppose u satisfies Laplace's equation inside the hemisphere $0 \leq r < 1$, $0 \leq \varphi \leq \pi/2$. If $u(r, \pi/2) = 0$ on $0 \leq r < 1$ and $u(1, \varphi) = 1$ on $0 \leq \varphi \leq \pi/2$, show that

$$u(r, \varphi) = \sum_{n=0}^{\infty} (-1)^n \left(\frac{4n+3}{2n+2} \right) \frac{(2n)!}{2^{2n}(n!)^2} r^{2n+1} P_{2n+1}(\cos \varphi)$$

inside the hemisphere.

4.33 Suppose u satisfies Laplace's equation inside the hemisphere $0 \leq r < R$, $0 \leq \varphi \leq \pi/2$. If $u_\varphi(r, \pi/2) = 0$ on $0 \leq r < R$, and $u(R, \varphi) = f(\varphi)$ on $0 \leq \varphi \leq \pi/2$, show that

$$u(r, \varphi) = \sum_{n=0}^{\infty} c_{2n} \left(\frac{r}{R} \right)^{2n} P_{2n}(\cos \varphi),$$

$$c_{2n} = (4n+1) \int_0^{\pi/2} f(\varphi) P_{2n}(\cos \varphi) \sin \varphi \, d\varphi.$$

5
Bessel Functions

We start by presenting the gamma function and some of its properties. This function is used to define the Bessel functions, hence its relevance to the subject of this chapter.

5.1 The Gamma Function

The *gamma function* is defined for all $x > 0$ by the improper integral

$$\Gamma(x) = \int_0^\infty e^{-t} t^{x-1} dt. \tag{5.1}$$

This integral converges for all positive values of x, therefore it clearly represents a continuous function on $(0, \infty)$. In fact we can show that Γ is of class C^∞ on $(0, \infty)$ (see Exercise 5.1). Integrating by parts, we have

$$\begin{aligned}\Gamma(x+1) &= \int_0^\infty e^{-t} t^x dt = -e^{-t} t^x \big|_0^\infty + x \int_0^\infty e^{-t} t^{x-1} dt \\ &= x \int_0^\infty e^{-t} t^{x-1} dt,\end{aligned}$$

which gives the characteristic recursion relation of the gamma function

$$\Gamma(x+1) = x\Gamma(x), \qquad x > 0. \tag{5.2}$$

If $x = n$ is a positive integer, then

$$\begin{aligned}\Gamma(n+1) &= n\Gamma(n) \\ &= n(n-1)\Gamma(n-1) \\ &\vdots \\ &= n!\Gamma(1).\end{aligned}$$

With

$$\Gamma(1) = \int_0^\infty e^{-t} dt = 1,$$

we have

$$\Gamma(n+1) = n!$$

which means that the gamma function Γ is an extension of the factorial mapping $n \mapsto (n-1)!$ from $\mathbb{N}$ to $(0, \infty)$.

Equation (5.2) implies

$$\Gamma(x) = \frac{\Gamma(x+1)}{x},$$

where the right-hand side can be extended to $(-1, 0) \cup (0, \infty)$. Since

$$\lim_{x \to 0} \Gamma(x+1) = \Gamma(1) = 1,$$

the gamma function has a simple pole at $x = 0$. Using the relation (5.2) repeatedly, we see that

$$\begin{aligned}\Gamma(x) &= \frac{\Gamma(x+1)}{x} \\ &= \frac{\Gamma(x+2)}{x(x+1)} \\ &\vdots \\ &= \frac{\Gamma(x+n)}{x(x+1)\cdots(x+n-1)}, \qquad n \in \mathbb{N}.\end{aligned}$$

This allows us to extend Γ from $(0, \infty)$ to $\mathbb{R}$, except for the integers $0, -1, -2, \ldots$, where Γ has simple poles (see Figure 5.1). Because

$$\Gamma(x) > \int_1^\infty e^{-t} t^{x-1} dt \geq \int_1^\infty e^{-t} dt = e^{-1}$$

for all $x \geq 1$, and $\Gamma(x+1) = x\Gamma(x)$, we also conclude that $\Gamma(x)$ tends to ∞ as $x \to \infty$.

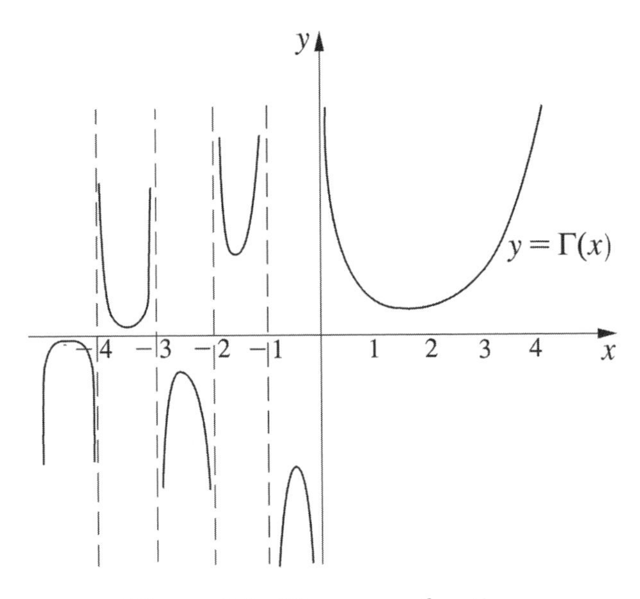

Figure 5.1 The gamma function.

EXERCISES

5.1 Prove that the gamma function, as defined by (5.1), belongs to $C^\infty(0,\infty)$.

5.2 Prove that $\Gamma(1/2) = \sqrt{\pi}$.

5.3 Prove that

$$\Gamma(n+1/2) = \frac{(2n)!\sqrt{\pi}}{n!2^{2n}}, \qquad n \in \mathbb{N}_0.$$

5.4 The *beta function* is defined by

$$\beta(x,y) = \int_0^1 t^{x-1}(1-t)^{y-1}dt, \qquad x>0, y>0.$$

(a) Use the transformation $u = t/(1-t)$ to obtain

$$\beta(x,y) = \int_0^\infty \frac{u^{x-1}}{(1+u)^{x+y}}du.$$

(b) Prove that, for any $s>0$,

$$\Gamma(z) = s^z \int_0^\infty e^{-st}t^{z-1}dt.$$

(c) With $s = u+1$ and $z = x+y$, show that

$$\frac{1}{(u+1)^{x+y}} = \frac{1}{\Gamma(x+y)} \int_0^\infty e^{-(u+1)t} t^{x+y-1} dt,$$

then use (a) to obtain the relation

$$\beta(x,y) = \frac{\Gamma(x)\Gamma(y)}{\Gamma(x+y)}.$$

5.5 Prove that

$$2^{2x} \frac{\Gamma(x)\Gamma(x+\frac{1}{2})}{\Gamma(2x)} = 2\sqrt{\pi}.$$

5.6 The *error function* on $\mathbb{R}$ is defined by the integral

$$\operatorname{erf}(x) = \frac{2}{\sqrt{\pi}} \int_0^x e^{-t^2} dt.$$

Sketch the graph of $\operatorname{erf}(x)$, and prove the following properties of the function:

(a) $\operatorname{erf}(-x) = -\operatorname{erf}(x)$.

(b) $\lim_{x\to\pm\infty} \operatorname{erf}(x) = \pm 1$.

(c) $\operatorname{erf}(x)$ is analytic in $\mathbb{R}$.

5.2 Bessel Functions of the First Kind

The differential equation

$$x^2 y'' + xy' + (x^2 - \nu^2)y = 0, \tag{5.3}$$

where ν is a nonnegative parameter, comes up in some situations where separation of variables is used to solve partial differential equations, as we show later in this chapter. It is called *Bessel's equation*, and we show that it is another example of an SL eigenvalue equation which generates certain special functions, called Bessel functions, in much the same way that the orthogonal polynomials of Chapter 4 were obtained. The main difference is that Bessel functions are not polynomials, and their orthogonality property is somewhat different.

Equation (5.3) has a singular point at $x = 0$, so we cannot expand the solution in a power series about that point. Instead, we use a method due to Georg Frobenius (1849–1917) to construct a solution in terms of real powers (not necessarily integers) of x. The method is based on the premise that every equation of the form

$$y'' + \frac{q(x)}{x}y' + \frac{r(x)}{x^2}y = 0,$$

where the functions q and r are analytic at $x = 0$, has a solution of the form

$$y(x) = x^t \sum_{k=0}^{\infty} c_k x^k = x^t(c_0 + c_1 x + c_2 x^2 + \cdots), \tag{5.4}$$

in which t is a real (or complex) number and the constant c_0 is nonzero [12]. Clearly t can always be chosen so that $c_0 \neq 0$. The expression (5.4) becomes a power series when t is a nonnegative integer.

Substituting the expression (5.4) into Equation (5.3), we obtain

$$\sum_{k=0}^{\infty}(k+t)(k+t-1)c_k x^{k+t} + \sum_{k=0}^{\infty}(k+t)c_k x^{k+t} + \sum_{k=0}^{\infty} c_k x^{k+t+2} - \nu^2 \sum_{k=0}^{\infty} c_k x^{k+t} = 0,$$

or

$$\sum_{k=0}^{\infty}(k+t)^2 c_k x^{k+t} + \sum_{k=0}^{\infty} c_k x^{k+t+2} - \nu^2 \sum_{k=0}^{\infty} c_k x^{k+t} = 0.$$

Collecting the coefficients of the powers x^t, x^{t+1}, x^{t+2}, $\ldots$, x^{t+j}, we obtain the following equations

$$t^2 c_0 - \nu^2 c_0 = 0 \tag{5.5}$$

$$(t+1)^2 c_1 - \nu^2 c_1 = 0 \tag{5.6}$$

$$(t+2)^2 c_2 - \nu^2 c_2 + c_0 = 0$$

$$\vdots$$

$$(t+j)^2 c_j - \nu^2 c_j + c_{j-2} = 0. \tag{5.7}$$

From Equation (5.5) we conclude that $t = \pm\nu$.

Assuming, to begin with, that $t = \nu$, Equation (5.6) becomes

$$(\nu+1)^2 c_1 - \nu^2 c_1 = (2\nu+1)c_1 = 0.$$

Since $2\nu + 1 \geq 1$, this implies $c_1 = 0$. Now Equation (5.7) yields

$$[(\nu+j)^2 - \nu^2]c_j + c_{j-2} = j(j+2\nu)c_j + c_{j-2} = 0,$$

and therefore

$$c_j = -\frac{1}{j(j+2\nu)}c_{j-2}. \tag{5.8}$$

Because $c_1 = 0$, it follows that $c_j = 0$ for all odd values of j, and we can assume that $j = 2m$, where m is a positive integer. The recursion relation (5.8) now takes the form

$$c_{2m} = -\frac{1}{2m(2m+2\nu)}c_{2m-2} = -\frac{1}{2^2 m(\nu+m)}c_{2m-2}, \qquad m \in \mathbb{N},$$

which allows us to express $c_2, c_4, c_6, \ldots$ in terms of the arbitrary constant c_0:

$$\begin{aligned}
c_2 &= -\frac{1}{2^2(\nu+1)}c_0 \\
c_4 &= -\frac{1}{2^2 2(\nu+2)}c_2 = \frac{1}{2^4 2!(\nu+1)(\nu+2)}c_0 \\
c_6 &= -\frac{1}{2^6 3!(\nu+1)(\nu+2)(\nu+3)}c_0 \\
&\vdots \\
c_{2m} &= \frac{(-1)^m}{2^{2m} m!(\nu+1)(\nu+2)\cdots(\nu+m)}c_0.
\end{aligned} \tag{5.9}$$

The resulting solution of Bessel's equation is therefore the formal series

$$x^\nu \sum_{m=0}^{\infty} c_{2m} x^{2m}. \tag{5.10}$$

By choosing

$$c_0 = \frac{1}{2^\nu \Gamma(\nu+1)} \tag{5.11}$$

the coefficients (5.9) are given by

$$c_{2m} = \frac{(-1)^m}{2^{\nu+2m} m! \Gamma(\nu+m+1)}.$$

The resulting series

$$x^\nu \sum_{m=0}^{\infty} \frac{(-1)^m}{2^{\nu+2m} m! \Gamma(v+m+1)} x^{2m}$$

is called Bessel's function of the first kind of order ν, and is denoted $J_\nu(x)$. Thus

$$J_\nu(x) = \left(\frac{x}{2}\right)^\nu \sum_{m=0}^{\infty} \frac{(-1)^m}{m! \Gamma(m+\nu+1)} \left(\frac{x}{2}\right)^{2m}, \tag{5.12}$$

and it is a simple matter to verify that the power series

$$\sum_{m=0}^{\infty} \frac{(-1)^m}{2^{2m} m! \Gamma(m+\nu+1)} x^{2m}$$

converges on $\mathbb{R}$ by the ratio test. With $\nu \geq 0$ the power x^ν is well defined when x is positive, hence the Bessel function $J_\nu(x)$ is well defined by (5.12) on $(0, \infty)$. Since

$$\lim_{x \to 0^+} J_\nu(x) = \begin{cases} 1, & \nu = 0 \\ 0, & \nu > 0, \end{cases}$$

the function J_ν may be extended as a continuous function to $[0, \infty)$ by the definition $J_\nu(0) = \lim_{x \to 0^+} J_\nu(x)$ for all $\nu \geq 0$.

Now if we set $t = -\nu < 0$ in (5.4), that is, if we change the sign of ν in (5.12), then

$$J_{-\nu}(x) = \left(\frac{x}{2}\right)^{-\nu} \sum_{m=0}^{\infty} \frac{(-1)^m}{m!\Gamma(m - \nu + 1)} \left(\frac{x}{2}\right)^{2m}, \qquad x > 0, \tag{5.13}$$

remains a solution of Bessel's equation, because the equation is invariant under such a change of sign. But it will not necessarily be bounded at $x = 0$, as we show in the next theorem.

Theorem 5.1

The Bessel functions J_ν and $J_{-\nu}$ are linearly independent if, and only if, ν is not an integer.

Proof

If $\nu = n \in \mathbb{N}_0$, then

$$J_{-n}(x) = \left(\frac{x}{2}\right)^{-n} \sum_{m=0}^{\infty} \frac{(-1)^m}{m!\Gamma(m - n + 1)} \left(\frac{x}{2}\right)^{2m}.$$

But because $1/\Gamma(m - n + 1) = 0$ for all $m - n + 1 \leq 0$, the terms in which $m = 0, 1, \ldots, n - 1$ all vanish, and we end up with

$$\begin{aligned} J_{-n}(x) &= \left(\frac{x}{2}\right)^{-n} \sum_{m=n}^{\infty} \frac{(-1)^m}{m!\Gamma(m - n + 1)} \left(\frac{x}{2}\right)^{2m} \\ &= \left(\frac{x}{2}\right)^{-n} \sum_{m=0}^{\infty} \frac{(-1)^{m+n}}{(m+n)!\Gamma(m + 1)} \left(\frac{x}{2}\right)^{2m+2n} \\ &= (-1)^n \left(\frac{x}{2}\right)^{n} \sum_{m=0}^{\infty} \frac{(-1)^m}{m!\Gamma(m + n + 1)} \left(\frac{x}{2}\right)^{2m} \\ &= (-1)^n J_n(x). \end{aligned}$$

Now suppose $\nu > 0,\ \nu \notin \mathbb{N}_0$, and let

$$aJ_\nu(x) + bJ_{-\nu}(x) = 0. \tag{5.14}$$

Taking the limit as $t \to 0$ in this equation, we have $\lim_{x\to 0^+} J_\nu(x) = 0$ whereas $\lim_{x\to 0^+} |J_{-\nu}| = \infty$ because the first term in the series (5.13), which is

$$\frac{1}{\Gamma(1-\nu)}\left(\frac{x}{2}\right)^{-\nu},$$

dominates all the other terms and tends to $\pm\infty$. Thus the equality (5.14) cannot hold on $(0,\infty)$ unless $b = 0$, in which case $a = 0$ as well. □

Based on this theorem we therefore conclude that, when ν is not an integer, the general solution of Bessel's equation on $(0,\infty)$ is given by $y(x) = c_1 J_\nu(x) + c_2 J_{-\nu}(x)$. The general solution when ν is an integer will have to await the definition of Bessel's function of the second kind in Section 5.3.

In the following example we prove the first of several identities involving the Bessel functions.

Example 5.2

$$\frac{d}{dx}[x^{-\nu}J_\nu(x)] = -x^{-\nu}J_{\nu+1}(x), \qquad x > 0, \nu \geq 0. \tag{5.15}$$

Proof

$x^{-\nu}J_\nu(x)$ is a power series, so it can be differentiated term by term:

$$\begin{aligned}
\frac{d}{dx}[x^{-\nu}J_\nu(x)] &= \frac{d}{dx}\sum_{m=0}^{\infty}\frac{(-1)^m}{2^{2m+\nu}m!\Gamma(m+\nu+1)}x^{2m} \\
&= \sum_{m=1}^{\infty}\frac{(-1)^m 2m}{2^{2m+\nu}m!\Gamma(m+\nu+1)}x^{2m-1} \\
&= -x^{-\nu}\sum_{m=0}^{\infty}\frac{(-1)^m}{2^{2m+\nu+1}m!\Gamma(m+\nu+2)}x^{2m+\nu+1} \\
&= -x^{-\nu}J_{\nu+1}.
\end{aligned}$$

□

Bessel's functions of integral order, given by

$$J_n(x) = \sum_{m=0}^{\infty} \frac{(-1)^m}{m!(m+n)!} \left(\frac{x}{2}\right)^{2m+n}, \qquad n \in \mathbb{N}_0, \tag{5.16}$$

are analytic in $(0, \infty)$, and have analytic extensions to $\mathbb{R}$ as even or odd functions, depending on whether n is even or odd. We have already established, in Example 2.13, that J_0 has an infinite set of isolated zeros in $(0, \infty)$ which accumulate at ∞. We can arrange these in an increasing sequence

$$\xi_{01} < \xi_{02} < \xi_{03} < \cdots$$

such that $\xi_{0k} \to \infty \text{ as } k \to \infty$. Using mathematical induction, we can show that the same is also true of J_n for any positive integer n: Suppose that the zeros of J_m, for any positive integer m, is an increasing sequence $(\xi_{mk} : k \in \mathbb{N})$ in $(0, \infty)$ which tends to ∞. The function $x^{-m}J_m(x)$ vanishes at any pair of consecutive zeros, say ξ_{mk} and $\xi_{m\ k+1}$, therefore it follows from Rolle's theorem that there is at least one point between ξ_{mk} and $\xi_{m\ k+1}$ where the derivative of $x^{-m}J_m(x)$ vanishes. In view of the identity (5.15),

$$J_{m+1}(x) = -x^m \frac{d}{dx}[x^{-m}J_m(x)],$$

hence $J_{m+1}(x)$ has at least one zero between ξ_{mk} and $\xi_{m\ k+1}$. Thus we have proved the following.

Theorem 5.3

For any $n \in \mathbb{N}_0$, the equation $J_n(x) = 0$ has an infinite number of positive roots, which form an increasing sequence

$$\xi_{n1} < \xi_{n2} < \xi_{n3} < \cdots$$

such that $\xi_{nk} \to \infty$ as $k \to \infty$.

The first two Bessel functions of integral order are

$$\begin{aligned}
J_0(x) &= \sum_{m=0}^{\infty} \frac{(-1)^m}{(m!)^2} \left(\frac{x}{2}\right)^{2m} \\
&= 1 - \frac{x^2}{2^2(1!)^2} + \frac{x^4}{2^4(2!)^2} - \frac{x^6}{2^6(3!)^2} + \cdots, \\
J_1(x) &= \sum_{m=0}^{\infty} \frac{(-1)^m}{m!(m+1)!} \left(\frac{x}{2}\right)^{2m+1} \\
&= \frac{x}{2} - \frac{x^3}{2^3 1! 2!} + \frac{x^5}{2^5 2! 3!} - \frac{x^7}{2^7 3! 4!} + \cdots.
\end{aligned}$$

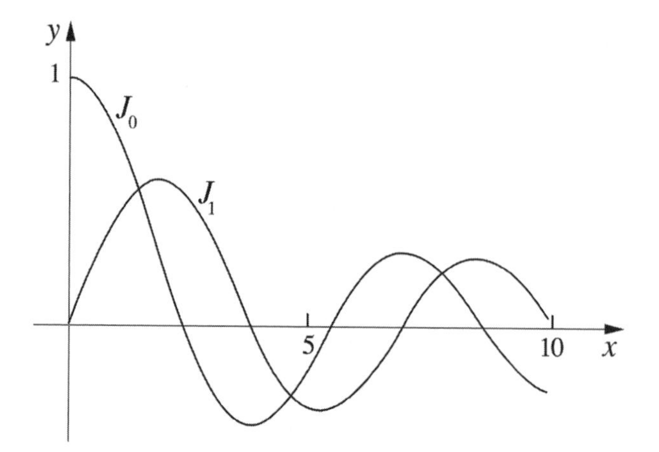

Figure 5.2 Bessel functions J_0 and J_1.

The similarities between these two expansions on the one hand, and those of the cosine and sine functions, are quite striking. The graphs of $J_0(x)$ and $J_1(x)$ in Figure 5.2 also exhibit many of the properties of $\cos x$ and $\sin x$, respectively, such as the behaviour near $x = 0$ and the interlacing of their zeros. When we set $\nu = 0$ in the identity (5.15) we obtain the relation $J_0'(x) = -J_1(x)$, which corresponds to the familiar relation $(\cos)'x = -\sin x$. But, unlike the situation with trigonometric functions, the distribution of the zeros of J_n is not uniform in general (see, however, Exercise 5.8) and the amplitude of the function decreases with increasing x (Exercise 5.32).

In the next example we prove another important relation between J_0 and J_1, one which we have occasion to resort to later in this chapter.

Example 5.4

$$\int_0^x tJ_0(t)dt = xJ_1(x) \qquad \text{for all } x > 0.$$

Proof

$$\begin{aligned}\int_0^x tJ_0(t)dt &= \int_0^x \sum_{m=0}^{\infty} \frac{(-1)^m}{(m!)^2 2^{2m}} t^{2m+1} dt \\ &= \sum_{m=0}^{\infty} \frac{(-1)^m x^{2m+2}}{(m!)^2 (2m+2) 2^{2m}} \\ &= x \sum_{m=0}^{\infty} \frac{(-1)^m}{m!(m+1)!} \left(\frac{x}{2}\right)^{2m+1} = xJ_1(x).\end{aligned}$$

□

EXERCISES

5.7 Verify that the power series which represents $x^{-\nu}J_\nu(x)$ converges on $\mathbb{R}$ for every $\nu \geq 0$.

5.8 Prove that

$$J_{1/2}(x) = \sqrt{\frac{2}{\pi x}}\sin x, \qquad J_{-1/2}(x) = \sqrt{\frac{2}{\pi x}}\cos x,$$

and sketch these two functions.

5.9 Prove that $xJ_\nu'(x) = \nu J_\nu(x) - xJ_{\nu+1}(x)$, and hence the identity of Example 5.2.

5.10 Use Exercises 5.8 and 5.9 to prove that

$$J_{3/2}(x) = \sqrt{\frac{2}{\pi x}}\left(\frac{\sin x}{x} - \cos x\right).$$

5.11 Prove the identity $[x^\nu J_\nu(x)]' = x^\nu J_{\nu-1}(x)$, and hence conclude that

$$J_{-3/2}(x) = -\sqrt{\frac{2}{\pi x}}\left(\frac{\cos x}{x} + \sin x\right).$$

5.12 Use the identities of Example 5.2 and Exercise 5.11 to establish

$$J_\nu'(x) = \frac{1}{2}[J_{\nu-1}(x) - J_{\nu+1}(x)].$$

5.13 Prove that

$$J_{\nu+1}(x) + J_{\nu-1}(x) = \frac{2\nu}{x}J_\nu(x).$$

5.14 Derive the following relations:

(a) $\int_0^x t^2 J_1(t)dt = 2xJ_1(x) - x^2 J_0(x)$.

(b) $\int_0^x J_3(t)dt = 1 - J_2(x) - 2J_1(x)/x$.

5.15 Use the identities $J_0'(x) = -J_1(x)$ and $[xJ_1(x)]' = xJ_0(x)$ to prove that

$$\int_0^x t^n J_0(t)dt = x^n J_1(x) + (n-1)x^{n-1}J_0(x) - (n-1)^2\int_0^x t^{n-2}J_0(t)dt.$$

5.16 Verify that the Wronskian $W(x) = W(J_\nu, J_{-\nu})$, where $\nu \notin \mathbb{N}_0$, satisfies the equation $xW' + W = 0$, and thereby prove that

$$W(x) = -\frac{2}{\Gamma(\nu)\Gamma(1-\nu)x}.$$

Using the result of Exercise 5.4(c), and evaluating the integral expression for $\beta(\nu, 1-\nu)$ by contour integration, it can be shown that

$$W(x) = \frac{-2\sin\nu\pi}{\pi x}$$

(also see [8]).

5.3 Bessel Functions of the Second Kind

In view of Theorem 5.1, it is natural to ask what the general solution of Bessel's equation looks like when ν is an integer n. There are several ways we can define a second solution to Bessel's equation which is independent of J_n (see, for example, Exercise 5.17). The more common approach is to define Bessel's function of the second kind of order ν by

$$Y_\nu(x) = \begin{cases} \dfrac{1}{\sin\nu\pi}[J_\nu(x)\cos\nu\pi - J_{-\nu}(x)], & \nu \neq 0,1,2,\ldots \\ \lim_{\nu\to n} Y_\nu, & n = 0,1,2,\ldots. \end{cases}$$

In this connection, observe the following points:

(i) For noninteger values of ν, Y_ν is a linear combination of J_ν and $J_{-\nu}$. Because $J_{-\nu}$ is linearly independent of J_ν, so is Y_ν.

(ii) When $\nu = n$, the above definition gives the indeterminate form $0/0$. By L'Hôpital's rule, using the differentiability properties of power series,

$$\begin{aligned} Y_n(x) &= \frac{1}{\pi}\left[\left.\frac{\partial J_\nu(x)}{\partial\nu}\right|_{\nu=n} - (-1)^n \left.\frac{\partial J_{-\nu}(x)}{\partial\nu}\right|_{\nu=n}\right] \\ &= \frac{2}{\pi}\left(\log\frac{x}{2}+\gamma\right)J_n(x) - \frac{1}{\pi}\left(\frac{x}{2}\right)^n \sum_{m=0}^{\infty}\frac{(-1)^m(h_m+h_{n+m})}{m!(n+m)!}\left(\frac{x}{2}\right)^{2m} \\ &\quad -\frac{1}{\pi}\left(\frac{x}{2}\right)^{-n}\sum_{m=0}^{n-1}\frac{(n-m-1)!}{m!}\left(\frac{x}{2}\right)^{2m}, \qquad x>0, \end{aligned}$$

where

$$h_0 = 0,\ h_m = 1 + \frac{1}{2} + \frac{1}{3} + \cdots + \frac{1}{m},$$

and

$$\gamma = \lim_{m\to\infty}(h_m - \log m) = 0.577215\cdots$$

is called Euler's constant. Note that the last sum in the expression for Y_n vanishes when $n = 0$, and that the presence of the term $\log x J_n(x)$ implies Y_n is linearly independent of J_n. That the passage to the limit as $\nu \to n$ preserves Y_n as a solution of Bessel's equation is due to the continuity of Bessel's equation and Y_ν with respect to ν. For more details on the computations which lead to the representation of Y_n given above, the reader is referred to [17], the classical reference on Bessel functions.

The asymptotic behaviour of $Y_n(x)$ as $x \to 0$ is given by

$$Y_n(x) \sim \begin{cases} \dfrac{2}{\pi}\log\dfrac{x}{2}, & n = 0 \\ -\dfrac{(n-1)!}{\pi}\left(\dfrac{x}{2}\right)^{-n}, & n \in \mathbb{N}, \end{cases} \tag{5.17}$$

where $f(x) \sim g(x)$ as $x \to c$ means $f(x)/g(x) \to 1$ as $x \to c$. Thus $Y_n(x)$ is unbounded in the neighborhood of $x = 0$. As $x \to 0$,

$$Y_0(x) \sim \frac{2}{\pi}\log x, \qquad Y_1(x) \sim -\frac{2}{\pi}\frac{1}{x}, \qquad Y_2(x) \sim \frac{4}{\pi}\frac{1}{x^2}, \quad \cdots.$$

In view of Theorem 5.3 and the Sturm separation theorem, we now conclude that Y_n has an infinite sequence of zeros in $(0, \infty)$ which alternate with the zeros of J_n. Y_0 and Y_1 are shown in Figure 5.3.

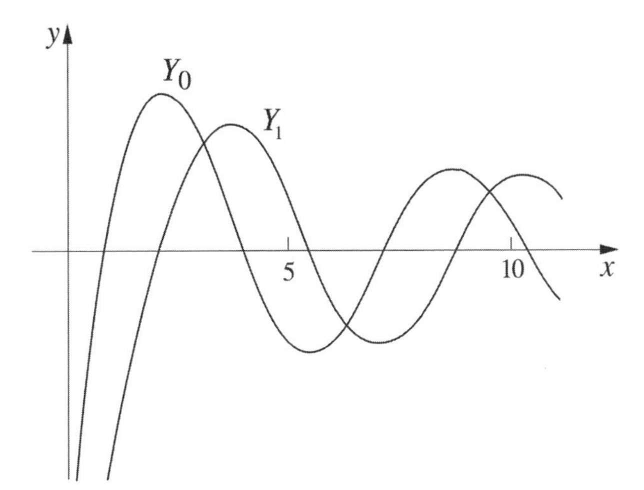

Figure 5.3 Bessel functions Y_0 and Y_1.

EXERCISES

5.17 Show that the function

$$y_n(x) = J_n(x) \int_c^x \frac{1}{t J_n^2(t)} dt$$

satisfies Bessel's equation and is linearly independent of $J_n(x)$.

5.18 Verify the asymptotic behaviour of Y_0 and Y_1 near $x = 0$ as expressed by (5.17).

5.19 Prove that

$$\frac{d}{dx}[x^\nu Y_\nu(x)] = x^\nu Y_{\nu-1}(x).$$

5.20 Prove that

$$\frac{d}{dx}[x^{-\nu} Y_\nu(x)] = -x^{-\nu} Y_{\nu+1}(x).$$

5.21 Prove that $Y_{-n}(x) = (-1)^n Y_n(x)$ for all $n \in \mathbb{N}_0$.

5.22 The modified Bessel function of the first kind I_ν is defined on $(0, \infty)$ by

$$I_\nu(x) = i^{-\nu} J_\nu(ix), \qquad \nu \geq 0,$$

where $i = \sqrt{-1}$. Show that I_ν satisfies the equation

$$x^2 y'' + x y' - (x^2 + \nu^2) y = 0. \tag{5.18}$$

5.23 Based on the definition of I_ν in Exercise 5.22, show that I_ν is a real function represented by the series

$$I_\nu(x) = \sum_{m=0}^{\infty} \frac{1}{m! \Gamma(m + \nu + 1)} \left(\frac{x}{2}\right)^{2m+\nu}.$$

5.24 Prove that $I_\nu(x) \neq 0$ for any $x > 0$, and that $I_{-n}(x) = I_n(x)$ for all $n \in \mathbb{N}$.

5.25 Show that the modified Bessel function of the second kind

$$K_\nu(x) = \frac{\pi}{2 \sin \nu\pi} [I_{-\nu}(x) - I_\nu(x)]$$

also satisfies Equation (5.18). I_0 and K_0 are shown in Figure 5.4.

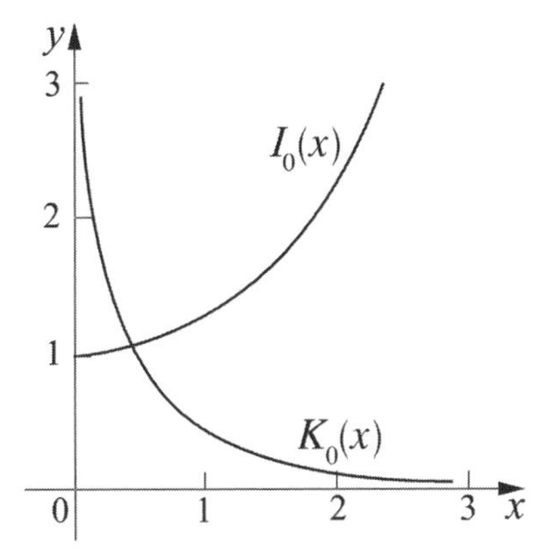

Figure 5.4 Modified Bessel functions I_0 and K_0.

5.4 Integral Forms of the Bessel Function J_n

We first prove that the generating function of J_n is

$$e^{x(z-1/z)/2} = \sum_{n=-\infty}^{\infty} J_n(x)z^n, \qquad z \neq 0. \tag{5.19}$$

This can be seen by noting that

$$e^{xz/2} = \sum_{j=0}^{\infty} \frac{z^j}{j!}\left(\frac{x}{2}\right)^j,$$
$$e^{-x/2z} = \sum_{k=0}^{\infty} \frac{(-1)^k}{k!z^k}\left(\frac{x}{2}\right)^k,$$

and that these two series are absolutely convergent for all x in $\mathbb{R}$, $z \neq 0$, hence their product is the double series

$$e^{x(z-1/z)/2} = \sum_{j=0}^{\infty}\sum_{k=0}^{\infty} \frac{(-1)^k}{j!k!}\left(\frac{x}{2}\right)^{j+k} z^{j-k}.$$

Setting $j-k=n$, and recalling that $1/(k+n)! = 1/\Gamma(k+n+1) = 0$ when $k+n<0$, we obtain

$$\begin{aligned} e^{x(z-1/z)/2} &= \sum_{n=-\infty}^{\infty}\left[\sum_{k=0}^{\infty} \frac{(-1)^k}{k!(k+n)!}\left(\frac{x}{2}\right)^{2k+n}\right] z^n \\ &= \sum_{n=-\infty}^{\infty} J_n(x)z^n, \end{aligned}$$

which proves (5.19). The substitution $z = e^{i\theta}$ now gives

$$\frac{1}{2}\left(z - \frac{1}{z}\right) = i \sin\theta,$$

hence

$$e^{ix\sin\theta} = \sum_{n=-\infty}^{\infty} J_n(x)e^{in\theta}. \tag{5.20}$$

The function $e^{ix\sin\theta}$ is periodic with period 2π and satisfies the conditions of Theorem 3.9, thus the right-hand side of (5.20) represents its Fourier series expansion in exponential form, and $J_n(x)$ are the Fourier coefficients in the expansion. Therefore

$$\begin{aligned} J_n(x) &= \frac{1}{2\pi}\int_{-\pi}^{\pi} e^{ix\sin\theta}e^{-in\theta}d\theta \\ &= \frac{1}{2\pi}\int_{-\pi}^{\pi} e^{i(x\sin\theta - n\theta)}d\theta \\ &= \frac{1}{2\pi}\int_{-\pi}^{\pi} \cos(x\sin\theta - n\theta)d\theta, \qquad \text{because } J_n(x) \text{ is real,} \\ &= \frac{1}{\pi}\int_{0}^{\pi} \cos(x\sin\theta - n\theta)d\theta, \qquad n \in \mathbb{N}_0, \end{aligned} \tag{5.21}$$

which is the principal integral representation of J_n. The formula (5.21) immediately gives an upper bound on J_n,

$$|J_n(x)| \leq \frac{1}{\pi}\int_0^{\pi} d\theta = 1 \qquad \text{for all } n \in \mathbb{N}_0,$$

a result we would not have been able to infer directly from the series definition. It also confirms that $J_0(0) = 1$ and $J_n(0) = 0$ for all $n \geq 1$.

Going back to Equation (5.20) and equating the real and the imaginary parts of both sides, we have

$$\cos(x\sin\theta) = \sum_{n=-\infty}^{\infty} J_n(x)\cos n\theta,$$

$$\sin(x\sin\theta) = \sum_{n=-\infty}^{\infty} J_n(x)\sin n\theta.$$

Because $J_{-n}(x) = (-1)^n J_n(x)$, this implies

$$\cos(x \sin\theta) = J_0(x) + 2\sum_{m=1}^{\infty} J_{2m}(x)\cos 2m\theta, \tag{5.22}$$

$$\sin(x \sin\theta) = 2\sum_{m=1}^{\infty} J_{2m-1}(x)\sin(2m-1)\theta. \tag{5.23}$$

Now Equations (5.22) and (5.23) are, respectively, the Fourier series expansions of the even function $\cos(x\sin\theta)$ and the odd function $\sin(x\sin\theta)$. Hence we arrive at the following pair of integral formulas,

$$J_{2m}(x) = \frac{1}{\pi}\int_0^{\pi} \cos(x\sin\theta)\cos 2m\theta \, d\theta, \qquad m \in \mathbb{N}_0, \tag{5.24}$$

$$J_{2m-1}(x) = \frac{1}{\pi}\int_0^{\pi} \sin(x\sin\theta)\sin(2m-1)\theta \, d\theta, \qquad m \in \mathbb{N}. \tag{5.25}$$

EXERCISES

5.26 Show that

$$J_n'(x) = \frac{1}{\pi}\int_0^{\pi} \sin\theta \sin(n\theta - x\sin\theta) d\theta,$$

then use induction, or Equation (5.20), to prove that

$$J_n^{(k)} = \frac{1}{\pi}\int_0^{\pi} \sin^k\theta \cos(n\theta - x\sin\theta - k\pi/2) d\theta, \qquad k \in \mathbb{N}.$$

5.27 Use the result of Exercise 5.26 to prove that $\left|J_n^{(k)}(x)\right| \leq 1$ for all $n, k \in \mathbb{N}_0$.

5.28 Prove that

(a) $J_0(x) + 2\sum_{m=1}^{\infty} J_{2m}(x) = 1$.

(b) $J_0(x) + 2\sum_{m=1}^{\infty}(-1)^m J_{2m}(x) = \cos x$.

(c) $2\sum_{m=0}^{\infty}(-1)^m J_{2m+1}(x) = \sin x$.

(d) $\sum_{m=1}^{\infty}(2m-1)J_{2m-1}(x) = x/2$.

5.29 Use Parseval's relation to prove the identity

$$J_0^2(x) + 2\sum_{n=1}^{\infty} J_n^2(x) = 1.$$

Observe that this implies $|J_0(x)| \leq 1$ and $|J_n(x)| \leq 1/\sqrt{2}$, $n \in \mathbb{N}$.

5.30 Use Equations (5.24) and (5.25) to conclude that

$$J_{2m}(x) = \frac{2}{\pi}\int_0^{\pi/2} \cos(x\sin\theta)\cos 2m\theta \, d\theta, \quad m \in \mathbb{N}_0,$$

$$J_{2m-1}(x) = \frac{2}{\pi}\int_0^{\pi/2} \sin(x\sin\theta)\sin(2m-1)\theta \, d\theta, \quad m \in \mathbb{N}.$$

5.31 Prove that $\lim_{n\to\infty} J_n(x) = 0$ for all $x \in \mathbb{R}$. Hint: Use Lemma 3.7.

5.32 Prove that $\lim_{x\to\infty} J_n(x) = 0$ for all $n \in \mathbb{N}_0$.

5.5 Orthogonality Properties

After division by x Bessel's equation takes the form

$$xy'' + y' + (x - \frac{\nu^2}{x})y = 0, \tag{5.26}$$

where the differential operator

$$L = \frac{d}{dx}\left(x\frac{d}{dx}\right) + x - \frac{\nu^2}{x}$$

is formally self-adjoint, with $p(x) = x$ and $r(x) = -\nu^2/x$ in the standard form (2.33). Comparison with Equation (2.34) shows that $\rho(x) = x$ is the weight function, but the eigenvalue parameter does not appear explicitly in Equation (5.26). We therefore introduce a parameter μ through the change of variables

$$x \mapsto \mu x, \; y(x) \mapsto y(\mu x) = u(x).$$

Differentiating with respect to x,

$$u'(x) = \mu y'(\mu x)$$
$$u''(x) = \mu^2 y''(\mu x).$$

Under this transformation, Equation (5.26) takes the form

$$xu'' + u' + \left(\mu^2 x - \frac{\nu^2}{x}\right) u = 0, \tag{5.27}$$

where the eigenvalue parameter is now $\lambda = \mu^2$. Equations (5.26) and (5.27) are equivalent provided $\mu \neq 0$.

If Equation (5.27) is given on the interval (a, b), where $0 \leq a < b < \infty$, then we can impose the homogeneous, separated boundary conditions

$$\alpha_1 u(a) + \alpha_2 u'(a) = 0, \qquad \beta_1 u(b) + \beta_2 u'(b) = 0,$$

to obtain a regular Sturm–Liouville eigenvalue problem. The eigenfunctions then have the form

$$c_\mu J_\nu(\mu x) + d_\mu Y_\nu(\mu x),$$

where μ, c_μ, and d_μ are chosen so that the boundary conditions are satisfied. The details are generally quite tedious, and we simplify things by taking $a = 0$.

Suppose, therefore, that Equation (5.27) is given on the interval $(0, b)$. Because $p(0) = 0$, no boundary condition is needed at $x = 0$, except that $\lim_{x \to 0^+} u(x)$ exist. At $x = b$ we have

$$\beta_1 u(b) + \beta_2 u'(b) = 0. \tag{5.28}$$

The pair of equations (5.27) and (5.28) now poses a singular SL eigenvalue problem and, based on an extension of the theory developed in Chapter 2, it has an orthogonal set of solutions which is complete in $\mathcal{L}_x^2(0, b)$. For the sake of simplicity, we restrict ν to the nonnegative integers. This allows us to focus on the main features of the theory without being bogged down in nonessential details. The assumption $\nu = n \in \mathbb{N}_0$ is also the most useful from the point of view of physical applications. The general solution of Equation (5.27) is then given by

$$u(x) = c_n J_n(\mu x) + d_n Y_n(\mu x).$$

The condition that $u(x)$ have a limit at $x = 0$ forces the coefficient of Y_n to vanish, and we are left with $J_n(\mu x)$ as the only admissible solution.

Let us start with the special case of Equation (5.28) when $\beta_2 = 0$; that is,

$$u(b) = 0. \tag{5.29}$$

Applying this condition to the solution $J_n(\mu x)$ gives

$$J_n(\mu b) = 0, \qquad n \in \mathbb{N}_0. \tag{5.30}$$

We have already determined in Theorem 5.3 that, for each n, the roots of Equation (5.30) in $(0,\infty)$ form an infinite increasing sequence which tends to ∞,

$$\xi_{n1} < \xi_{n2} < \xi_{n3} < \cdots .$$

The solutions of Equation (5.30) are therefore given by

$$\mu_k b = \xi_{nk},$$

and the eigenvalues of Equation (5.27) are

$$\lambda_k = \mu_k^2 = \left(\frac{\xi_{nk}}{b}\right)^2, \qquad k \in \mathbb{N}.$$

Note that the first zero $\xi_{n0} = 0$ of the function J_n, for $n \geq 1$, does not determine an eigenvalue because the corresponding solution is

$$J_n(\mu_0 x) = J_n(0) = 0 \quad \text{for all } n \in \mathbb{N},$$

which is not an eigenfunction. The sequence of eigenvalues of the system (5.27), (5.28) is therefore

$$0 < \lambda_1 = \mu_1^2 < \lambda_2 = \mu_2^2 < \lambda_3 = \mu_3^2 < \cdots ,$$

and its corresponding sequence of eigenfunctions is

$$J_n(\mu_1 x),\ J_n(\mu_2 x),\ J_n(\mu_3 x),\ \ldots .$$

For each $n \in \mathbb{N}_0$, the sequence $(J_n(\mu_k x) : k \in \mathbb{N})$ is necessarily orthogonal and complete in $\mathcal{L}_x^2(0,b)$. In other words

$$\begin{aligned} \left\langle J_n(\mu_j x), J_n(\mu_k x)\right\rangle_x &= \int_0^b J_n(\mu_j x) J_n(\mu_k x) x dx \\ &= 0 \ \text{ for all } j \neq k, \end{aligned} \tag{5.31}$$

and, for any $f \in \mathcal{L}_x^2(0,b)$ and any $n \in \mathbb{N}_0$, we can represent f by the Fourier-Bessel series

$$f(x) = \sum_{k=1}^{\infty} \frac{\langle f(x), J_n(\mu_k x)\rangle_x}{\|J_n(\mu_k x)\|_x^2} J_n(\mu_k x), \tag{5.32}$$

the latter equality being, of course, in $\mathcal{L}_x^2(0,b)$. If f is piecewise smooth on $(0,b)$ Equation(5.32) holds pointwise as well, provided $f(x)$ is defined as $\frac{1}{2}[f(x^+) + f(x^-)]$ at the points of discontinuity.

To verify the orthogonality relation (5.31) by direct computation is not a simple matter, but we can attempt to determine $\|J_n(\alpha_k x)\|$. Multiplying Equation (5.27) by $2xu'$,

$$2xu'(xu')' + (\mu^2x^2 - \nu^2)2uu' = 0$$
$$[(xu')^2]' + (\mu^2x^2 - \nu^2)(u^2)' = 0.$$

Integrating this last equation over $(0, b)$,

$$(xu')^2\Big|_0^b + \mu^2\left[x^2u^2\Big|_0^b - 2\int_0^b xu^2dx\right] - \nu^2u^2\Big|_0^b = 0,$$

$$\Rightarrow \quad \|u\|_x^2 = \frac{1}{2\mu^2}\left[(\mu xu)^2 + (xu')^2 - \nu^2u^2\right]\Big|_0^b.$$

With $\nu = n \in \mathbb{N}_0$, $u(x) = J_n(\mu x)$, and $u'(x) = \mu J_n'(\mu x)$, $\mu > 0$, we therefore have

$$\|J_n(\mu x)\|_x^2 = \int_0^b J_n^2(\mu x)xdx$$
$$= \frac{1}{2\mu^2}\left[(\mu^2x^2 - n^2)J_n^2(\mu x) + \mu^2x^2J_n'^2(\mu x)\right]\Big|_0^b.$$

Because $n^2J_n(0) = 0$ for all $n \in \mathbb{N}_0$,

$$\|J_n(\mu x)\|_x^2 = \frac{b^2}{2}\left[J_n'(\mu b)\right]^2 + \frac{\mu^2b^2 - n^2}{2\mu^2}J_n^2(\mu b). \tag{5.33}$$

When $\mu = \mu_k$ the last term drops out, by (5.30), and

$$\|J_n(\mu_k x)\|_x^2 = \frac{b^2}{2}[J_n'(\mu_k b)]^2.$$

Using the result of Exercise 5.9,

$$J_n'(\mu_k b) = \frac{1}{\mu_k b}[nJ_n(\mu_k b) - \mu_k bJ_{n+1}(\mu_k b)] = -J_{n+1}(\mu_k b),$$

we finally obtain

$$\|J_n(\mu_k x)\|_x^2 = \frac{b^2}{2}J_{n+1}^2(\mu_k b). \tag{5.34}$$

Example 5.5

To expand the function

$$f(x) = \begin{cases} 1, & 0 \le x < 2 \\ 0, & 2 < x \le 4 \end{cases}$$

in a Fourier–Bessel series under the condition $J_0(4\mu) = 0$, we first evaluate the coefficients in the series:

$$\begin{aligned}\langle f(x), J_0(\mu_k x)\rangle_x &= \int_0^4 f(x) J_0(\mu_k x) x dx \\ &= \int_0^2 J_0(\mu_k x) x dx \\ &= \frac{1}{\mu_k^2} \int_0^{2\mu_k} J_0(y) y dy \\ &= \frac{2}{\mu_k} J_1(2\mu_k),\end{aligned}$$

where we used the result of Example 5.4 in the last equality, $4\mu_k$ being the zeros of J_0. From the formula (5.34) we get

$$\|J_0(\mu_k x)\|_x^2 = 8 J_1^2(4\mu_k).$$

Now, by (5.32),

$$f(x) = \frac{1}{4} \sum_{k=1}^{\infty} \frac{J_1(2\mu_k)}{\mu_k J_1^2(4\mu_k)} J_0(\mu_k x), \qquad 0 < x < 4.$$

Observe that, by setting $x = 1$ in this equation, we arrive at the identity

$$\sum_{k=1}^{\infty} \frac{J_1(2\mu_k) J_0(\mu_k)}{\mu_k J_1^2(4\mu_k)} = 4,$$

because 1 is a point of continuity of f. At the point of discontinuity $x = 2$, we have

$$\sum_{k=1}^{\infty} \frac{J_1(2\mu_k) J_0(2\mu_k)}{\mu_k J_1^2(4\mu_k)} = 4 \frac{f(2^+) + f(2^-)}{2} = 2.$$

The more general condition (5.28)

$$\beta_1 u(b) + \beta_2 u'(b) = 0$$

can, in principle, be handled in a similar fashion, although it requires more work. If $\beta_1 \neq 0$ and $\beta_2 \neq 0$, we can assume, without loss of generality, that $\beta_2 = 1$. The solution $u(x) = J_n(\mu x)$ must then satisfy

$$\beta_1 J_n(\mu b) + \mu J_n'(\mu b) = 0. \tag{5.35}$$

The positive roots of this equation determine the eigenvalues μ_k^2 of the problem. The corresponding eigenfunctions are $u_k(x) = J_n(\mu_k x)$. The norm of each eigenfunction is determined by substituting into (5.33) to obtain

$$\begin{aligned} \|J_n(\mu_k x)\|_x^2 &= \frac{b^2}{2}\left[J_n'(\mu_k b)\right]^2 + \frac{\mu_k^2 b^2 - n^2}{2\mu_k^2} J_n^2(\mu_k b) \\ &= \frac{1}{2\mu_k^2}\left[\beta_1^2 b^2 + \mu_k^2 b^2 - n^2\right] J_n^2(\mu_k b). \end{aligned} \tag{5.36}$$

To solve Equation (5.35) for μ is, in general, not a simple matter. We should recall that J_n and J_n' cannot have a common zero, because the solution to the initial-value problem for Bessel's equation is unique, so one would most likely have to resort to numerical methods to determine μ_k. But if $\beta_1 = 0$, that is, if the boundary condition at b is

$$u'(b) = 0, \tag{5.37}$$

then the eigenvalues are determined by the roots of the equation $\mu J_n'(\mu b) = 0$. The first is $\mu_0 = 0$, corresponding to the eigenvalue

$$\lambda_0 = 0,$$

and the eigenfunction

$$u_0(x) = J_0(0) = 1.$$

Note that this function solves Equation (5.27) but not (5.26). As mentioned earlier, these two equations are not equivalent if $\mu = 0$.

With $n = 0$, the other values of μ_k must satisfy $J_0'(\mu_k b) = 0$ or, equivalently, $J_1(\mu_k b) = 0$. Thus the sequence

$$\mu_k = \xi_{1k}/b, \qquad k = 1, 2, 3, \ldots,$$

where ξ_{1k} are the positive zeros of J_1, yields the remaining eigenvalues of the problem,

$$\lambda_k = \frac{\xi_{1k}^2}{b^2}, \qquad k \in \mathbb{N},$$

corresponding to the eigenfunctions

$$u_k(x) = J_0(\mu_k x), \qquad k \in \mathbb{N}.$$

Under the boundary condition (5.37), if n is a positive integer, then 0 is not an eigenvalue, and we obtain the eigenvalues $\lambda_k = \mu_k^2$ by solving

$$J_n'(\mu_k b) = 0.$$

The corresponding eigenfunctions are then $u_k(x) = J_n(\mu_k x)$. In any case, whether n is 0 or positive, the norm of J_n is calculated from the formula (5.36) with $\beta_1 = 0$.

Example 5.6

In $\mathbb{R}^3$ the cylindrical coordinates (r, θ, z) are related to the Cartesian coordinates (x, y, z) by

$$x = r\cos\theta, \qquad y = r\sin\theta,$$
$$r = \sqrt{x^2 + y^2},$$

where $0 \le r < \infty$ and $-\pi < \theta \le \pi$. The Laplacian operator in cylindrical coordinates has the form

$$\Delta = \frac{\partial^2}{\partial r^2} + \frac{1}{r}\frac{\partial}{\partial r} + \frac{1}{r^2}\frac{\partial^2}{\partial \theta^2} + \frac{\partial^2}{\partial z^2}.$$

Given the cylindrical region

$$\Omega = \{(r, \theta, z) : 0 \le r < b, -\pi < \theta \le \pi, 0 < z < h\},$$

we seek a potential function u which satisfies Laplace's equation

$$\Delta u = u_{rr} + r^{-1}u_r + r^{-2}u_{\theta\theta} + u_{zz} = 0$$

in Ω, and assumes the following values on the boundary of Ω:

$$\begin{aligned} u(r,\theta,0) &= 0, & 0 \le r \le b, \quad & -\pi < \theta \le \pi, \\ u(b,\theta,z) &= 0, & -\pi < \theta \le \pi, \quad & 0 \le z \le h, \\ u(r,\theta,h) &= f(r), & 0 \le r \le b, \quad & -\pi < \theta \le \pi. \end{aligned}$$

This is often referred to as a Dirichlet problem in Ω.

The general Dirichlet problem is to determine the function u in a domain Ω which satisfies Laplace's equation $\Delta u = 0$ in Ω and $u = f$ on the boundary $\partial\Omega$, where f is a given function defined on $\partial\Omega$. When $\partial\Omega$ and f satisfy certain regularity conditions, the problem is known to have a unique solution in $C^2(\Omega)$ (see [13]).

Solution

Because u is independent of θ on the boundary, we can assume, by symmetry, that it is also independent of θ inside Ω. Using separation of variables, let

$$u(r, z) = v(r)w(z), \qquad 0 \le r \le b,$$

and substitute into Laplace's equation to obtain

$$v''(r)w(z) + r^{-1}v'(r)w(z) + v(r)w''(z) = 0.$$

This leads to the pair of equations

$$r^2 v''(r) + rv'(r) + \mu^2 r^2 v(r) = 0 \tag{5.38}$$

$$w''(z) - \mu^2(z) = 0, \tag{5.39}$$

where the separation constant was assumed to be $-\mu^2$. Equation (5.38) is Bessel's equation with $\nu = 0$, whose continuous solution in $0 \le r \le b$ is $J_0(\mu r)$. Under the boundary condition $v(b) = 0$, the eigenfunctions are $J_0(\mu_k r)$ corresponding to the eigenvalues $\lambda_k = \mu_k^2 = (\xi_{0k}/b)^2$. As before, ξ_{01}, ξ_{02}, $\xi_{03}, \ldots$ are the positive zeros of J_0.

The corresponding eigenfunctions of Equation (5.39) under the condition $w(0) = 0$ are $\sinh(\mu_k z)$. Hence the sequence of eigenfunctions of the original problem is

$$u_k(r, z) = J_0(\mu_k r)\sinh(\mu_k z), \qquad k \in \mathbb{N}.$$

Before applying the nonhomogeneous boundary condition at $z = h$, we form the more general solution

$$u(r, z) = \sum_{k=1}^{\infty} c_k u_k(r, z).$$

Now the condition $u(r, h) = f(r)$ implies

$$f(r) = \sum_{k=1}^{\infty} c_k \sinh(\mu_k h) J_0(\mu_k r), \tag{5.40}$$

which is the Fourier–Bessel series expansion of the function f, assuming of course that f is piecewise smooth on $(0, b)$. The coefficients in the expansion are given by

$$c_k \sinh(\mu_k h) = \frac{1}{\|J_0(\mu_k r)\|_r^2} \int_0^b f(r) J_0(\mu_k r) r \, dr,$$

and this completely determines c_k for all positive integers k. Bessel functions usually appear in the solution of Laplace's equation when cylindrical coordinates are used, hence they are often referred to as *cylinder functions.*

EXERCISES

5.33 Determine the Fourier–Bessel series expansion $\sum c_k J_0(\mu_k x)$ on $[0, b]$, where μ_k are the positive roots of the equation $J_0(\mu b) = 0$, for each of the following functions.

(a) $f(x) = 1$

(b) $f(x) = x$

(c) $f(x) = x^2$

(d) $f(x) = b^2 - x^2$

(e) $f(x) = \begin{cases} 1, & 0 < x < b/2 \\ 1/2, & x = b/2 \\ 0, & b/2 < x < b. \end{cases}$

5.34 Expand the function $f(x) = 1$ on $[0, 1]$ in terms of $J_0(\mu_k x)$, where μ_k are the nonnegative zeros of J_0'.

5.35 Expand $f(x) = x$ on $[0, 1]$ in terms of $J_1(\mu_k x)$, where μ_k are the positive zeros of J_1.

5.36 For any positive integer n, expand $f(x) = x^n$ on $[0, 1]$ in terms of $J_n(\mu_k x)$, where μ_k are the positive zeros of J_n'.

5.37 Determine the coefficients in the series $\sum c_k J_1(\mu_k x)$ which represents the function

$$f(x) = \begin{cases} x, & 0 \le x \le 1 \\ 0, & 1 < x \le 2 \end{cases}$$

on $[0, 2]$, where μ_k are the positive zeros of $J_1'(2\mu)$. Is the representation pointwise on $[0, 2]$?

5.38 Show that 0 is an eigenvalue of Bessel's equation (5.27) subject to the boundary condition (5.28) if, and only if, $\beta_1/\beta_2 = -\nu/b$, and that the corresponding eigenfunction is x^ν, where $\nu > 0$.

5.39 The heat equation on a circular flat plate is given in polar coordinates (r, θ) by

$$u_t = k\left(u_{rr} + \frac{1}{r}u_r + \frac{1}{r^2}u_{\theta\theta}\right).$$

Suppose that the temperature $u = u(r, t)$ does not depend on θ and that $0 \le r < 1$. If the edge of the plate is held at zero temperature for all $t > 0$, use separation of variables to show that the temperature on the plate is given by

$$u(r, t) = \sum_{n=1}^{\infty} c_n e^{-\mu_n^2 kt} J_0(\mu_n r) \quad \text{for all} r \in [0, 1), \qquad t > 0, \quad (5.41)$$

where μ_n are the positive zeros of J_0.

5.40 If the initial temperature on the plate in Exercise 5.39 is $u(r, 0) = f(r)$, prove that the Fourier–Bessel coefficients in (5.41) are given by

$$c_n = \frac{2}{J_1^2(\mu_n)} \int_0^1 f(r) J_0(\mu_n r) r dr.$$

5.41 A thin elastic circular membrane vibrates transversally according to the wave equation

$$u_{tt} = c^2 \left(u_{rr} + \frac{1}{r} u_r \right), \qquad 0 \leq r < R, t > 0.$$

If the boundary condition is $u(R, t) = 0$ for all $t > 0$, and the initial conditions are

$$u(r, 0) = f(r), \qquad u_t(r, 0) = g(r), \qquad 0 \leq r < R,$$

determine the form of the bounded solution $u(r, t)$ in terms of J_n for all $r \in [0, R)$ and $t > 0$.

6
The Fourier Transformation

The underlying theme of the previous chapters was the Sturm–Liouville theory. The last three chapters show how the eigenfunctions of various SL problems serve as bases for $\mathcal{L}^2$, either through conventional Fourier series or its generalized version. In this chapter we introduce the Fourier integral as a limiting case of the classical Fourier series, and show how it serves, under certain conditions, as a method for representing nonperiodic functions on $\mathbb{R}$ where the series approach does not apply. This chapter and the next are therefore concerned with extending the theory of Fourier series to nonperiodic functions.

6.1 The Fourier Transform

Suppose $f : \mathbb{R} \to \mathbb{C}$ is an $\mathcal{L}^2$ function. Its restriction to $(-l, l)$ clearly lies in $\mathcal{L}^2(-l, l)$ for any $l > 0$. On the interval $(-l, l)$ we can always represent f by the Fourier series

$$f(x) = \sum_{n=-\infty}^{\infty} c_n e^{in\pi x/l}, \tag{6.1}$$

$$c_n = \frac{1}{2l} \int_{-l}^{l} f(x) e^{-in\pi x/l} dx, \qquad n \in \mathbb{Z}. \tag{6.2}$$

Let $\Delta\xi = \pi/l$ and $\xi_n = n\Delta\xi = n\pi/l$. The pair of equations (6.1) and (6.2) then take the form

$$f(x) = \frac{1}{2\pi}\sum_{-\infty}^{\infty} C(\xi_n)e^{i\xi_n x}\Delta\xi, \tag{6.3}$$

$$C(\xi_n) = 2lc_n = \int_{-l}^{l} f(x)e^{-i\xi_n x}dx. \tag{6.4}$$

If we now let $l \to \infty$, that is, if we allow the period $(-l, l)$ to increase to $\mathbb{R}$ so that f loses its periodicity, then the discrete variable ξ_n will behave more as a real variable ξ, and the formula (6.4) will tend to the form

$$C(\xi) = \int_{-\infty}^{\infty} f(x)e^{-i\xi x}dx. \tag{6.5}$$

The right-hand side of (6.3), on the other hand, looks very much like a Riemann sum which, in the limit as $l \to \infty$, approaches the integral

$$f(x) = \frac{1}{2\pi}\int_{-\infty}^{\infty} C(\xi)e^{ix\xi}d\xi. \tag{6.6}$$

Thus the Fourier coefficients c_n are transformed to the function $C(\xi)$, the Fourier transform of f, and the Fourier series (6.1), which represents f on $(-l, l)$, is replaced by the Fourier integral (6.6) which, presumably, represents the function f on $(-\infty, \infty)$.

The procedure described above is, of course, not intended to be a "proof" of the validity of the formulas (6.5) and (6.6). The integral in (6.5) may not even exist. It is meant to be a plausible argument for motivating the definition (to follow) of the Fourier transform (6.5), which can then be used to represent the (nonperiodic) function f by the integral (6.6), in much the same way that the Fourier series were used in Chapter 3 to represent periodic functions.

For any real interval I, we use the symbol $\mathcal{L}^1(I)$ to denote the set of functions $f : I \to \mathbb{C}$ such that

$$\int_I |f(x)|\, dx < \infty.$$

Thus, if I is a bounded interval, any integrable function on I, and in particular any piecewise continuous function, belongs to $\mathcal{L}^1(I)$. Furthermore,

$$x^\alpha \in \mathcal{L}^1(0, 1) \Leftrightarrow \alpha > -1,$$

$$x^\alpha \in \mathcal{L}^1(1, \infty) \Leftrightarrow \alpha < -1.$$

If I is unbounded, a function may be integrable (in the improper sense) without belonging to $\mathcal{L}^1(I)$, such as $\sin x/x$ over $(0,\infty)$ (see Exercise 1.44). That is why we refer to $\mathcal{L}^1(I)$ functions as absolutely integrable functions on I in order to avoid this confusion when I is unbounded. As with $\mathcal{L}^2(I)$, it is a simple matter to check that $\mathcal{L}^1(I)$ is also a linear space.

Definition 6.1

For any $f \in \mathcal{L}^1(\mathbb{R})$ we define the *Fourier transform* of f as the function $\hat{f} : \mathbb{R} \to \mathbb{C}$ defined by the improper integral

$$\hat{f}(\xi) = \int_{-\infty}^{\infty} f(x)e^{-i\xi x}dx. \tag{6.7}$$

We also use the symbol $\mathcal{F}(f)$ instead of $\hat{f}$ to denote the Fourier transform of f.

Because $\left|e^{i\xi x}\right| = 1$ we clearly have

$$\left|\hat{f}(\xi)\right| \leq \int_{-\infty}^{\infty} |f(x)|\, dx < \infty;$$

that is, $\hat{f}$ is a bounded function on $\mathbb{R}$. By the linearity of the integral,

$$\mathcal{F}(c_1 f_1 + c_2 f_2) = c_1\mathcal{F}(f_1) + c_2\mathcal{F}(f_2)$$

for all $c_1, c_2 \in \mathbb{C}$ and all $f_1, f_2 \in \mathcal{L}^1(\mathbb{R})$, which just means that the Fourier transformation $\mathcal{F} : f \to \hat{f}$ is linear.

Example 6.2

For any positive constant a, let

$$f_a(x) = \begin{cases} 1, & |x| \leq a \\ 0, & |x| > a. \end{cases}$$

Then

$$\hat{f}_a(\xi) = \int_{-a}^{a} e^{-i\xi x}dx = \frac{1}{-i\xi}\left(e^{-i\xi a} - e^{i\xi a}\right) = \frac{2}{\xi}\sin a\xi,$$

as shown in Figure 6.1. Note that $\lim_{a\to\infty} \hat{f}_a(\xi)$ does not exist and that $f(x) = 1$ does not lie in $\mathcal{L}^1(\mathbb{R})$.

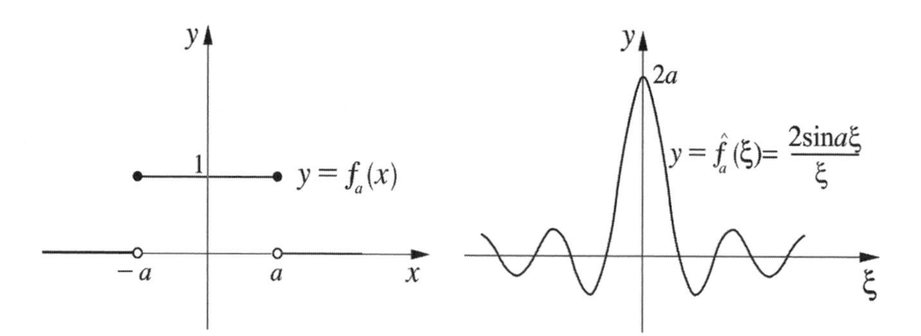

Figure 6.1

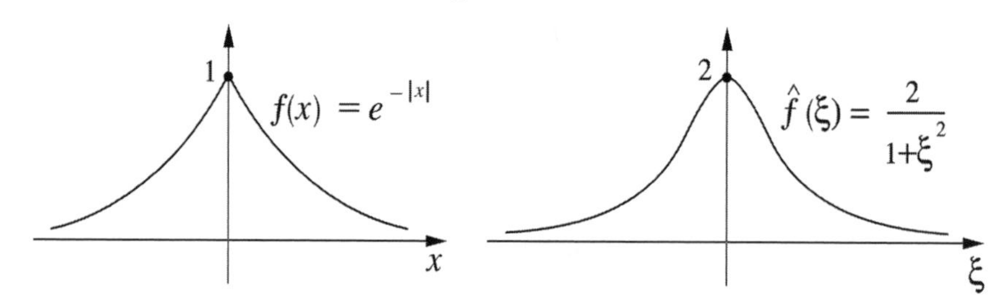

Figure 6.2

Example 6.3

In the case of the function $f(x) = e^{-|x|}$ we have (see Figure 6.2)

$$\begin{aligned}\hat{f}(\xi) &= \int_{-\infty}^{0} e^{x} e^{-i\xi x} dx + \int_{0}^{\infty} e^{-x} e^{-i\xi x} dx \\ &= \frac{1}{1-i\xi} + \frac{1}{1+i\xi} \\ &= \frac{2}{1+\xi^2}.\end{aligned}$$

When $|f|$ is integrable over $\mathbb{R}$ (i.e., when $f \in \mathcal{L}^1(\mathbb{R})$) we have seen that its Fourier transform $\hat{f}$ is bounded, but we can also prove that $\hat{f}$ is continuous. This result relies on a well-known theorem of real analysis, the Lebesgue dominated convergence theorem, which states the following.

Theorem 6.4

Let $(f_n : n \in \mathbb{N})$ be a sequence of functions in $\mathcal{L}^1(I)$, where I is a real interval, and suppose $f_n \to f$ pointwise on I. If there is a positive function $g \in \mathcal{L}^1(I)$ such that

$$|f_n(x)| \leq g(x) \qquad \text{for all } x \in I, \qquad n \in \mathbb{N},$$

then $f \in \mathcal{L}^1(I)$ and

$$\lim_{n\to\infty} \int_I f_n(x)dx = \int_I f(x)dx.$$

The proof of this theorem may be found in [1] or [14], where it is presented within the context of Lebesgue integration, but it is equally valid when the integrals are interpreted as Riemann integrals. Note that the interval I is not assumed to be finite.

To prove that $\hat{f}$ is continuous, let ξ be any real number and suppose (ξ_n) is a sequence which converges to ξ. Since

$$\left|\hat{f}(\xi_n) - \hat{f}(\xi)\right| \leq \int_{-\infty}^{\infty} \left|e^{-i\xi_n x} - e^{-i\xi x}\right| |f(x)|\, dx,$$

$$\left|e^{-i\xi_n x} - e^{-i\xi x}\right| |f(x)| \leq 2\, |f(x)| \in \mathcal{L}^1(\mathbb{R}),$$

and

$$\lim_{n\to\infty} \left|e^{-i\xi_n x} - e^{-i\xi x}\right| = 0,$$

Theorem 6.4 implies

$$\begin{aligned}
\lim_{n\to\infty} \left|\hat{f}(\xi_n) - \hat{f}(\xi)\right| &\leq \lim_{n\to\infty} \int_{-\infty}^{\infty} \left|e^{-i\xi_n x} - e^{-i\xi x}\right| |f(x)|\, dx \\
&= \int_{-\infty}^{\infty} \lim_{n\to\infty} \left|e^{-i\xi_n x} - e^{-i\xi x}\right| |f(x)|\, dx \\
&= 0.
\end{aligned}$$

In order to study the behaviour of $\hat{f}(\xi)$ as $|\xi| \to \infty$, we need the following result, often referred to as the Riemann–Lebesgue lemma.

Lemma 6.5

Let f be a piecewise smooth function on $\mathbb{R}$.
(i) If $[a, b]$ is a bounded interval, then

$$\lim_{|\xi|\to\infty} \int_a^b f(x)e^{i\xi x}dx = 0.$$

(ii) If $f \in \mathcal{L}^1(\mathbb{R})$, then

$$\lim_{|\xi|\to\infty} \int_{-\infty}^{\infty} f(x)e^{i\xi x}dx = 0.$$

Proof

(i) Let $x_1, x_2, \ldots, x_n$ be the points of discontinuity of f and f' in (a, b), arranged in increasing order, and let $a = x_0$ and $b = x_{n+1}$. We then have

$$\int_a^b f(x)e^{i\xi x}dx = \sum_{k=0}^{k=n} \int_{x_k}^{x_{k+1}} f(x)e^{i\xi x}dx,$$

and it suffices to prove that

$$\lim_{|\xi|\to\infty} \int_{x_k}^{x_{k+1}} f(x)e^{i\xi x}dx = 0 \qquad \text{for all } k.$$

Integrating by parts,

$$\int_{x_k}^{x_{k+1}} f(x)e^{i\xi x}dx = \frac{1}{i\xi} f(x)e^{i\xi x}\Big|_{x_k}^{x_{k+1}} - \frac{1}{i\xi} \int_{x_k}^{x_{k+1}} f'(x)e^{i\xi x}dx,$$

and the right-hand side of this equation clearly tends to 0 as $\xi \to \pm\infty$.

(ii) Let ε be any positive number. Because $|f|$ is integrable on $(-\infty, \infty)$, we know that there is a positive number L such that

$$\left| \int_{-\infty}^{\infty} f(x)e^{i\xi x}dx - \int_{-L}^{L} f(x)e^{i\xi x}dx \right| \leq \int_{|x|>L} |f(x)|\, dx < \frac{\varepsilon}{2}.$$

But, from part (i), we also know that there is a positive number K such that

$$\left| \int_{-L}^{L} f(x)e^{i\xi x}dx \right| < \frac{\varepsilon}{2} \qquad \text{for all } |\xi| > K.$$

Therefore, if $|\xi| > K$, then $\left| \int_{-\infty}^{\infty} f(x)e^{i\xi x}dx \right| < \varepsilon$. $\square$

We have therefore proved the following theorem.

Theorem 6.6

For any $f \in \mathcal{L}^1(\mathbb{R})$, the Fourier transform

$$\hat{f}(x) = \int_{-\infty}^{\infty} f(x)e^{-i\xi x}dx$$

is a bounded continuous function on $\mathbb{R}$. If, furthermore, f is piecewise smooth, then

$$\lim_{|\xi|\to\infty} \hat{f}(\xi) = 0. \tag{6.8}$$

Remark 6.7

1. Lemma 6.5 clearly remains valid if $e^{i\xi x}$ is replaced by either $\cos \xi x$ or $\sin \xi x$.
2. The Riemann–Lebesgue lemma actually holds under the weaker condition that $|f|$ is merely integrable, but the proof requires a little more work (see [16] for example). In any case we do not need this more general result, because Lemma 6.5 is used in situations (such as the proof of Theorem 6.10) where f is assumed to be piecewise smooth.
3. In view of the above remark, Equation (6.8) holds for any $f \in \mathcal{L}^1(\mathbb{R})$ without assuming piecewise smoothness.

We indicated in our heuristic introduction to this chapter that the Fourier transform $\hat{f}$, denoted C in (6.5), plays the role of the Fourier coefficients of the periodic function f in the limit as the function loses its periodicity. Hence the asymptotic behaviour $\hat{f}(\xi) \to 0$ as $\xi \to \pm\infty$ is in line with the behaviour of the Fourier coefficients c_n when $n \to \pm\infty$. But although Theorem 6.6 states some basic properties of $\hat{f}$, it says nothing about its role in the representation of f as suggested by (6.6), namely the validity of the formula

$$f(x) = \frac{1}{2\pi} \int_{-\infty}^{\infty} \hat{f}(\xi) e^{ix\xi} d\xi. \tag{6.9}$$

The right-hand side of this equation is called the Fourier integral of f. This integral may not exist, even if the function $\hat{f}(\xi)$ tends to 0 as $|\xi| \to \infty$, unless it tends to 0 fast enough. Furthermore, even if the Fourier integral exists, the equality (6.9) may not hold pointwise in $\mathbb{R}$. This is the subject of the next section.

EXERCISES

6.1 Determine the Fourier transform of each of the following functions.

(a) $f(x) = \begin{cases} 1 - |x|, & |x| \leq 1 \\ 0, & |x| > 1. \end{cases}$

(b) $f(x) = \begin{cases} \cos x, & |x| \leq \pi \\ 0, & |x| > \pi. \end{cases}$

(c) $f(x) = \begin{cases} 1, & 0 \leq x \leq 1 \\ 0, & \text{otherwise.} \end{cases}$

6.2 Given any $f : I \to \mathbb{C}$, prove that

(a) If I is bounded and $f \in \mathcal{L}^2(I)$, then $f \in \mathcal{L}^1(I)$,

(b) If f is bounded and $f \in \mathcal{L}^1(I)$, then $f \in \mathcal{L}^2(I)$.

6.3 Let $\varphi : I \times J \to \mathbb{C}$, where I and J are real intervals, and suppose that, for each $x \in I$, $\varphi(x, \cdot)$ is a continuous function on J. If $\varphi(\cdot, \xi)$ is integrable on I for each $\xi \in J$, and there is a positive function $g \in \mathcal{L}^1(I)$ such that $|\varphi(x, \xi)| \leq g(x)$ for all $x \in I$ and $\xi \in J$, use the dominated convergence theorem to prove that the function $F(\xi) = \int_I \varphi(x, \xi) dx$ is continuous on J.

6.4 Under the hypothesis of Exercise 6.3, if $\varphi(x, \cdot)$ is piecewise continuous on J, prove that F is also piecewise continuous on J.

6.5 If the function $\varphi(x, \cdot)$ in Exercise 6.3 is differentiable on J and $|\varphi_\xi(x, \cdot)| \leq h(x)$ for some $h \in \mathcal{L}^1(I)$, prove that F is differentiable and that its derivative $F'(\xi)$ equals $\int_I \varphi_\xi(x, \xi) dx$. If φ_ξ is continuous on J, prove that F' is also continuous on J. Hint: For any $\xi \in J$, $\psi_n(x, \xi) = (\xi_n - \xi)^{-1}[\varphi(x, \xi_n) - \varphi(x, \xi)] \to \varphi_\xi(x, \xi)$ as $\xi_n \to \xi$. Use the mean value theorem to conclude that $|\psi_n(x, \xi)| \leq |h(x)|$ on $I \times J$.

6.6 Using the equality

$$\int_0^\infty e^{-\xi x} dx = \frac{1}{\xi} \quad \text{for all } \xi > 0,$$

show that, for any positive number a,

$$\int_0^\infty x^n e^{-\xi x} dx = \frac{n!}{\xi^{n+1}}, \quad \xi > a, \quad n \in \mathbb{N}.$$

Note that if we set $\xi = 1$ in this last equation, we arrive at the relation $n! = \Gamma(n + 1)$.

6.7 Use Exercises 6.3 and 6.5 to deduce that the gamma function

$$\Gamma(\xi) = \int_0^\infty e^{-x} x^{\xi - 1} dx$$

is continuous on $[a, b]$ for any $0 < a < b < \infty$, and that its nth derivative

$$\Gamma^{(n)}(\xi) = \int_0^\infty e^{-x} \frac{d^n}{d\xi^n} \left(x^{\xi - 1}\right) dx$$

is continuous on $[a, b]$. Conclude from this that $\Gamma \in C^\infty(0, \infty)$.

6.8 Use Lemma 6.5 to evaluate the following limits, where D_n is the Dirichlet kernel defined in Section 3.2.

(a) $\lim_{n\to\infty} \displaystyle\int_{-\pi/2}^{\pi/2} D_n(x) dx$

(b) $\lim_{n\to\infty} \displaystyle\int_0^{\pi/2} D_n(x) dx$

(c) $\lim_{n\to\infty} \int_{\pi/6}^{\pi/2} D_n(x)dx$

6.9 Let f and g be piecewise smooth functions on (a,b), and suppose that $x_1, \ldots, x_n$ are their points of discontinuity. Prove the following generalization of the formula for integration by parts.

$$\int_a^b f(x)g'(x)dx = f(b^-)g(b^-) - f(a^+)g(a^+)$$

$$- \int_a^b f'(x)g(x)dx + \sum_{k=1}^{n-1} [f(x_{k+1}^-)g(x_{k+1}^-) - f(x_k^+)g(x_k^+)].$$

6.2 The Fourier Integral

The main result of this section is Theorem 6.10, which establishes the inversion formula for the Fourier transform. The proof of the theorem relies on evaluating the improper integral

$$\int_0^\infty \frac{\sin x}{x} dx,$$

which is known as Dirichlet's integral. To show that this integral exists, we write

$$\int_0^\infty \frac{\sin x}{x} dx = \int_0^1 \frac{\sin x}{x} dx + \lim_{b\to\infty} \int_1^b \frac{\sin x}{x} dx. \tag{6.10}$$

Because the function $\sin x/x$ is continuous and bounded on $(0,1]$, where it satisfies $0 \le \sin x/x \le 1$, the first integral on the right-hand side of (6.10) exists. Using integration by parts in the second integral,

$$\int_1^b \frac{\sin x}{x} dx = \cos 1 - \frac{\cos b}{b} - \int_1^b \frac{\cos x}{x^2} dx,$$

and noting that

$$\begin{aligned} \left| \int_1^b \frac{\cos x}{x^2} dx \right| &\le \int_1^b \left| \frac{\cos x}{x^2} \right| dx \\ &\le \int_1^b \frac{1}{x^2} dx \\ &\le 1 - \frac{1}{b}, \end{aligned}$$

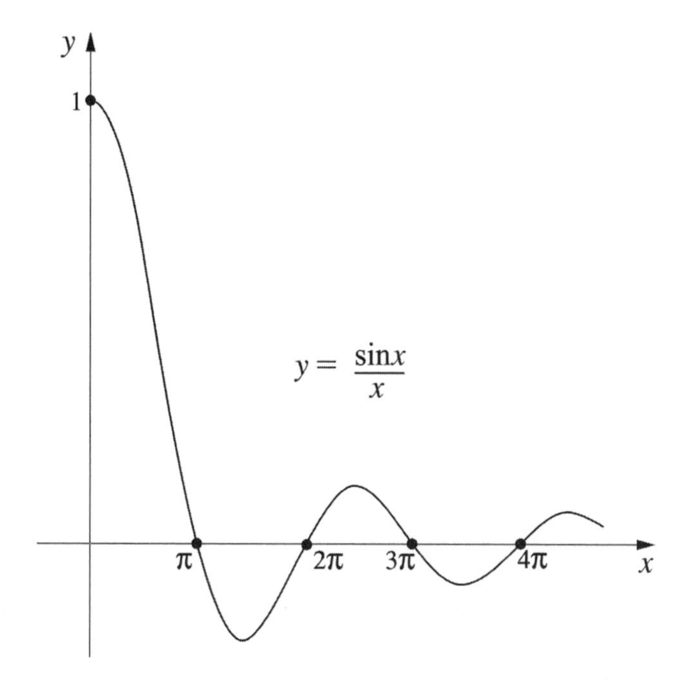

Figure 6.3

we see that $\lim_{b\to\infty}\int_1^b(\cos x/x^2)dx$ exists. Hence the integral $\int_1^\infty(\sin x/x)dx$ is convergent (see Exercise 1.44 for an alternative approach). The integrand $\sin x/x$ is shown graphically in Figure 6.3.

Now that we know Dirichlet's integral exists, it remains to determine its value.

Lemma 6.8

$$\int_0^\infty \frac{\sin x}{x}dx = \frac{\pi}{2}. \tag{6.11}$$

Proof

Define the function

$$f(x) = \begin{cases} \dfrac{1}{x} - \dfrac{1}{2\sin\frac{1}{2}x}, & 0 < x \le \pi \\ 0, & x = 0. \end{cases}$$

It is a simple matter to check that f and its derivative are both continuous on $[0, \pi]$. By Lemma 6.5,

$$\lim_{\xi\to\infty} \int_0^\pi f(x) \sin \xi x \, dx = 0. \tag{6.12}$$

Therefore

$$\begin{aligned}
\int_0^\infty \frac{\sin x}{x} dx &= \lim_{\xi\to\infty} \int_0^{\pi\xi} \frac{\sin x}{x} dx \\
&= \lim_{\xi\to\infty} \int_0^{\pi} \frac{\sin \xi x}{x} dx \\
&= \lim_{\xi\to\infty} \frac{1}{2} \int_0^{\pi} \frac{\sin \xi x}{\sin \frac{1}{2}x} dx \quad \text{by (6.12)} \\
&= \lim_{n\to\infty} \frac{1}{2} \int_0^{\pi} \frac{\sin (n+\frac{1}{2})x}{\sin \frac{1}{2}x} dx.
\end{aligned}$$

Going back to the definition of the Dirichlet kernel

$$D_n(x) = \frac{1}{2\pi} \sum_{k=-n}^{n} e^{ikx},$$

and using Lemma 3.8 to write

$$D_n(x) = \frac{1}{2\pi} \frac{\sin (n+\frac{1}{2})x}{\sin \frac{1}{2}x},$$

we conclude that

$$\int_0^\infty \frac{\sin x}{x} dx = \lim_{n\to\infty} \pi \int_0^\pi D_n(x) dx = \frac{\pi}{2},$$

by Equation (3.17). □

Although the function

$$\frac{\sin \xi x}{x}$$

is continuous with respect to both $x > 0$ and $\xi \in \mathbb{R}$, the function defined by the improper integral

$$K(\xi) = \int_0^\infty \frac{\sin \xi x}{x} dx$$

is not continuous at $\xi = 0$, for

$$K(\xi) = \begin{cases} \pi/2, & \xi > 0 \\ 0, & \xi = 0 \\ -\pi/2, & \xi < 0. \end{cases}$$

This implies that the function $|\sin \xi x/x|$ is not dominated (bounded) by an $\mathcal{L}^1(0,\infty)$ function, which clearly follows from the fact that $|\sin x\xi/x|$ is not integrable on $(0,\infty)$ (see Exercise 1.44). The situation we have here is analogous to the convergence of the Fourier series

$$\sum_{n=1}^{\infty} \frac{\sin n\xi}{n}$$

to a discontinuous function because its convergence is not uniform. Based on this analogy, the improper integral

$$F(\xi) = \int_a^{\infty} \varphi(x,\xi)dx$$

is said to be uniformly convergent on the interval I if, given any $\varepsilon > 0$, there is a number $N > a$ such that

$$b > N \;\Rightarrow\; \left| F(\xi) - \int_a^b \varphi(x,\xi)dx \right| = \left| \int_b^{\infty} \varphi(x,\xi)dx \right| < \varepsilon \qquad \text{for all } \xi \in I.$$

The number N depends on ε and is independent of ξ.

Corresponding to the Weierstrass M-test for uniform convergence of series, we have the following test for the uniform convergence of improper integrals. We leave the proof as an exercise.

Lemma 6.9

Let $\varphi : [a,\infty) \times I \to \mathbb{C}$, and suppose that there is a function $g \in \mathcal{L}^1(a,\infty)$ such that $|\varphi(x,\xi)| \leq g(x)$ for all ξ in the interval I. Then the integral $\int_a^{\infty} \varphi(x,\xi)dx$ is uniformly convergent on I.

If a function $\varphi(x,\xi)$ satisfies the conditions of Lemma 6.9 with $I = [\alpha,\beta]$ and if, in addition, $\varphi(x,\cdot)$ is continuous on $[\alpha,\beta]$ for each $x \in [a,\infty)$, then the function

$$F(\xi) = \int_a^{\infty} \varphi(x,\xi)dx$$

is also continuous on $[\alpha,\beta]$ (Exercise 6.3), and satisfies

$$\int_{\alpha}^{\beta} F(\xi)d\xi = \int_a^{\infty} \int_{\alpha}^{\beta} \varphi(x,\xi)d\xi dx. \tag{6.13}$$

This follows from the observation that the uniform convergence

$$\int_a^b \varphi(x,\xi)dx \to F(\xi) \quad \text{as} \quad b \to \infty$$

implies that, for every $\varepsilon > 0$, there is an $N > 0$ such that

$$b \geq N \Rightarrow \left| F(\xi) - \int_a^b \varphi(x,\xi)dx \right| \leq \int_b^\infty g(x)dx < \varepsilon$$

$$\Rightarrow \left| \int_\alpha^\beta F(\xi)d\xi - \int_a^b \int_\alpha^\beta \varphi(x,\xi)d\xi dx \right| = \left| \int_\alpha^\beta F(\xi)d\xi - \int_\alpha^\beta \int_a^b \varphi(x,\xi)dxd\xi \right|$$
$$\leq \varepsilon(\beta - \alpha).$$

In other words, under the hypothesis of Lemma 6.9, we can change the order of integration in the double integral

$$\int_a^\infty \int_\alpha^\beta \varphi(x,\xi)d\xi dx = \int_\alpha^\beta \int_a^\infty \varphi(x,\xi)dxd\xi.$$

This equality remains valid in the limit as $\beta \to \infty$ provided F is integrable on $[\alpha, \infty)$.

Pushing the analogy with Fourier series to its logical conclusion, we should expect to be able to reconstruct an $\mathcal{L}^1$ function f solely from knowledge of its transform $\hat{f}$, in the same way that a periodic function is determined (according to Theorem 3.9) by its Fourier coefficients. Based on Remark 3.10 and Equation (6.6), we are tempted to write

$$f(x) = \frac{1}{2\pi} \int_{-\infty}^\infty \hat{f}(\xi) e^{ix\xi} d\xi,$$

assuming that x is a point of continuity of f, and we would not be too far off the mark. The only problem is that, although continuous and bounded, $\left|\hat{f}\right|$ may not be integrable on $(-\infty, \infty)$, so that the above integral may not converge. Some treatments introduce a damping function, such as $e^{-\varepsilon^2\xi^2/2}$, into the integrand to force convergence, and then take the limit of the resulting integral as $\varepsilon \to 0$. Here we introduce a cut-off function, already suggested by the Fourier series representation

$$f(x) = \lim_{N\to\infty} \sum_{n=-N}^{N} c_n e^{inx}, \tag{6.14}$$

which is further elaborated on in Remark 6.11.

Corresponding to Theorem 3.9 for Fourier series, we have the following fundamental theorem for representing an $\mathcal{L}^1$ function by a Fourier integral.

Theorem 6.10

Let f be a piecewise smooth function in $\mathcal{L}^1(\mathbb{R})$. If

$$\hat{f}(\xi) = \int_{-\infty}^{\infty} f(x)e^{-i\xi x}dx, \qquad \xi \in \mathbb{R}, \tag{6.15}$$

then

$$\lim_{L\to\infty} \frac{1}{2\pi}\int_{-L}^{L} \hat{f}(\xi)e^{ix\xi}d\xi = \frac{1}{2}[f(x^+) + f(x^-)] \qquad \text{for all } x \in \mathbb{R}. \tag{6.16}$$

Before attempting to prove this theorem, it is worthwhile to consider some important observations on the meaning and implications of Equation (6.16).

Remark 6.11

1. The limit

$$\lim_{L\to\infty} \frac{1}{2\pi}\int_{-L}^{L} \hat{f}(\xi)e^{ix\xi}d\xi \tag{6.17}$$

is the Cauchy principal value of the improper integral

$$\frac{1}{2\pi}\int_{-\infty}^{\infty} \hat{f}(\xi)e^{ix\xi}d\xi = \lim_{\substack{a\to-\infty \\ b\to\infty}} \frac{1}{2\pi}\int_{a}^{b} \hat{f}(\xi)e^{ix\xi}d\xi. \tag{6.18}$$

The restricted limit (6.17), as is well known, may exist even when the more general limit (6.18) does not. If $\hat{f}$ lies in $\mathcal{L}^1(\mathbb{R})$ then, of course, the two limits are equal.

2. If f is defined by

$$f(x) = \frac{1}{2}[f(x^+) + f(x^-)] \tag{6.19}$$

at every point of discontinuity x, then (6.16) becomes

$$f(x) = \lim_{L\to\infty} \frac{1}{2\pi}\int_{-L}^{L} \hat{f}(\xi)e^{ix\xi}dx = \mathcal{F}^{-1}(\hat{f}),$$

the inverse Fourier transform of $\hat{f}$, or the Fourier integral of $\hat{f}$.

3. When $\hat{f} \in \mathcal{L}^1(\mathbb{R})$ and (6.19) holds, Equations (6.15) and (6.16) define the transform pair

$$\hat{f}(\xi) = \mathcal{F}(f)(\xi) = \int_{-\infty}^{\infty} f(x)e^{-i\xi x}dx,$$

$$f(x) = \mathcal{F}^{-1}(\hat{f}) = \frac{1}{2\pi}\int_{-\infty}^{\infty} \hat{f}(\xi)e^{ix\xi}d\xi,$$

a representation which exhibits a high degree of symmetry. Some books adopt the definition

$$\hat{f}(\xi) = \frac{1}{\sqrt{2\pi}} \int_{-\infty}^{\infty} f(x) e^{-ix\xi} dx,$$

from which

$$f(x) = \frac{1}{\sqrt{2\pi}} \int_{-\infty}^{\infty} \hat{f}(\xi) e^{i\xi x} d\xi,$$

and thereby achieve complete symmetry. Our definition is a natural development of the notation used in Fourier series.
4. If $\hat{f} = 0$ then (6.16) implies $f = 0$, assuming of course that Equation (6.19) is valid. This means that the Fourier transformation $\mathcal{F}$, defined on the piecewise smooth functions in $\mathcal{L}^1(\mathbb{R})$, is injective.

Proof

$$\begin{aligned} \int_{-L}^{L} \hat{f}(\xi) e^{ix\xi} d\xi &= \int_{-L}^{L} \left[\int_{-\infty}^{\infty} f(y) e^{-i\xi y} dy \right] e^{ix\xi} d\xi \\ &= \int_{-\infty}^{\infty} \int_{-L}^{L} f(y) e^{i\xi(x-y)} d\xi dy, \end{aligned}$$

where we used Equation (6.13) to change the order of integration with respect to y and ξ, because the function

$$f(y) e^{i\xi(x-y)}$$

satisfies the conditions of Lemma 6.9. Integrating with respect to ξ, we obtain

$$\begin{aligned} \int_{-L}^{L} \hat{f}(\xi) e^{ix\xi} d\xi &= 2 \int_{-\infty}^{\infty} \frac{\sin L(x-y)}{x-y} f(y) dy \\ &= 2 \int_{-\infty}^{\infty} \frac{\sin L\eta}{\eta} f(x+\eta) d\eta. \end{aligned} \tag{6.20}$$

Suppose now that δ is an arbitrary positive number. As a function of η, $f(x+\eta)/\eta$ is piecewise smooth and its absolute value is integrable on $(-\infty, -\delta] \cup [\delta, \infty)$, so we can use Lemma 6.5 to conclude that

$$\lim_{L\to\infty} \int_{|\eta|\geq\delta} \frac{\sin L\eta}{\eta} f(x+\eta) d\eta = 0.$$

Taking the limits of both sides of Equation (6.20) as $L \to \infty$, we therefore have

$$\begin{aligned} \lim_{L\to\infty} \int_{-L}^{L} \hat{f}(\xi) e^{ix\xi} d\xi &= 2 \lim_{L\to\infty} \int_{-\delta}^{\delta} \frac{\sin L\eta}{\eta} f(x+\eta) d\eta \\ &= 2 \lim_{L\to\infty} \int_{0}^{\delta} \frac{\sin L\eta}{\eta} [f(x+\eta) + f(x-\eta)] d\eta. \end{aligned}$$

But

$$\lim_{L\to\infty}\int_0^\delta \frac{\sin L\eta}{\eta}f(x+\eta)d\eta = \lim_{L\to\infty}\int_0^\delta \sin L\eta \frac{f(x+\eta)-f(x^+)}{\eta}d\eta + \lim_{L\to\infty} f(x^+)\int_0^\delta \frac{\sin L\eta}{\eta}d\eta.$$

Now Lemma 6.5 implies

$$\lim_{L\to\infty}\int_0^\delta \sin L\eta \frac{f(x+\eta)-f(x^+)}{\eta}d\eta = 0,$$

and Lemma 6.8 gives

$$\lim_{L\to\infty}\int_0^\delta \frac{\sin L\eta}{\eta}d\eta = \int_0^\infty \frac{\sin x}{x}dx = \frac{\pi}{2},$$

hence

$$\lim_{L\to\infty}\int_0^\delta \frac{\sin L\eta}{\eta}f(x+\eta)d\eta = \frac{\pi}{2}f(x^+).$$

Similarly,

$$\lim_{L\to\infty}\int_0^\delta \frac{\sin L\eta}{\eta}f(x-\eta)d\eta = \frac{\pi}{2}f(x^-).$$

Consequently,

$$\lim_{L\to\infty}\int_{-L}^{L} \hat{f}(\xi)e^{ix\xi}d\xi = \pi[f(x^+)+f(x^-)],$$

and we obtain the desired equality after dividing by 2π. □

The similarity between Theorems 3.9 and 6.10 is quite striking, although it is somewhat obscured by the fact that the trigonometric form of the Fourier series was used in the statement and proof of Theorem 3.9, whereas Theorem 6.10 is expressed in terms of the exponential form of the Fourier transform and integral. The correspondence becomes clearer when the formulas in Theorem 3.9 are cast in the exponential form

$$c_n = \frac{1}{2\pi}\int_{-\pi}^{\pi} f(x)e^{-ix\xi}dx,$$

$$\lim_{N\to\infty}\sum_{n=-N}^{N} c_n e^{inx} = \frac{1}{2}[f(x^+)+f(x^-)].$$

On the other hand, the trigonometric form of Theorem 6.10 is easily derived from Euler's relation $e^{i\theta} = \cos\theta + i\sin\theta$. Suppose $f \in \mathcal{L}^1(\mathbb{R})$ is a real, piecewise smooth function which satisfies

$$f(x) = \frac{1}{2}[f(x^+)+f(x^-)]$$

at each point of discontinuity. Such a function has a Fourier transform

$$\begin{aligned}\hat{f}(\xi) &= \int_{-\infty}^{\infty} f(x)e^{-i\xi x}dx \\ &= \int_{-\infty}^{\infty} f(x)\cos \xi x dx - i\int_{-\infty}^{\infty} f(x)\sin \xi x\ dx \\ &= A(\xi) - iB(\xi), \end{aligned} \tag{6.21}$$

where

$$A(\xi) = \int_{-\infty}^{\infty} f(x)\cos \xi x\ dx, \tag{6.22}$$

$$B(\xi) = \int_{-\infty}^{\infty} f(\xi)\sin \xi x\ dx, \quad \xi \in \mathbb{R}, \tag{6.23}$$

are the Fourier cosine and the Fourier sine transforms of f, respectively.

From Theorem 6.10 we can now express $f(x)$ as

$$f(x) = \lim_{L\to\infty} \frac{1}{2\pi}\int_{-L}^{L} [A(\xi) - iB(\xi)]e^{ix\xi}d\xi.$$

Because $A(\xi)$ is even and $B(\xi)$ is odd, we have

$$\begin{aligned}f(x) &= \lim_{L\to\infty} \frac{1}{2\pi}\int_{-L}^{L} [A(\xi)\cos x\xi + B(\xi)\sin x\xi]d\xi \\ &= \lim_{L\to\infty} \frac{1}{\pi}\int_{0}^{L} [A(\xi)\cos x\xi + B(\xi)\sin x\xi]d\xi \\ &= \frac{1}{\pi}\int_{0}^{\infty} [A(\xi)\cos x\xi + B(\xi)\sin x\xi]d\xi, \quad x \in \mathbb{R}. \end{aligned} \tag{6.24}$$

If f is even, then (6.23) implies $B(\xi) = 0$ and the transform pair (6.22) and (6.24) take the form

$$A(\xi) = 2\int_{0}^{\infty} f(x)\cos \xi x\ dx, \tag{6.25}$$

$$f(x) = \frac{1}{\pi}\int_{0}^{\infty} A(\xi)\cos x\xi\ d\xi. \tag{6.26}$$

If f is odd, then $A(\xi) = 0$ and we obtain

$$B(\xi) = 2\int_{0}^{\infty} f(x)\sin \xi x\ dx, \tag{6.27}$$

$$f(x) = \frac{1}{\pi}\int_{0}^{\infty} B(x)\sin x\xi\ d\xi. \tag{6.28}$$

The attentive reader will not miss the analogy between these formulas and the corresponding formulas for a_n and b_n in the trigonometric Fourier series representation of a periodic function.

Example 6.12

Applying Theorem 6.10 to Example 6.2, we see that

$$\lim_{L\to\infty}\frac{1}{2\pi}\int_{-L}^{L}\frac{2}{\xi}\sin a\xi e^{ix\xi}d\xi = \lim_{L\to\infty}\frac{1}{\pi}\int_{-L}^{L}\frac{1}{\xi}\sin a\xi\cos x\xi\, d\xi$$

$$= \frac{2}{\pi}\int_{0}^{\infty}\frac{1}{\xi}\sin a\xi\cos x\xi\, d\xi$$

because $\sin a\xi/\xi$ is an even function and $\cos\xi x$ is the even (real) part of $e^{i\xi x}$. Therefore, recalling the definition of f_a, we arrive at the following interesting evaluation of the above integral,

$$\frac{2}{\pi}\int_{0}^{\infty}\frac{1}{\xi}\sin a\xi\cos x\xi\, d\xi = \begin{cases} 0, & x<-a \\ 1/2, & x=-a \\ 1, & -a<x<a \\ 1/2, & x=a \\ 0, & x>a. \end{cases}$$

The fact that the right-hand side of this equality is not a continuous function implies that the integral in the left-hand side does not converge uniformly.

Example 6.13

The function

$$f(x) = \begin{cases} \sin x, & |x|<\pi \\ 0, & |x|>\pi \end{cases}$$

is odd. Hence its cosine and sine transforms are, respectively,

$$A(\xi) = 0,$$

and

$$B(\xi) = 2\int_{0}^{\infty} f(x)\sin\xi x\, dx$$

$$= 2\int_{0}^{\pi}\sin x\sin\xi x\, dx$$

$$= \frac{1}{1-\xi}\sin(1-\xi)\pi - \frac{1}{1+\xi}\sin(1+\xi)\pi$$

$$= \frac{2\sin\pi\xi}{1-\xi^2}.$$

$f(x)$ and $B(\xi)$ are shown in Figure 6.4.

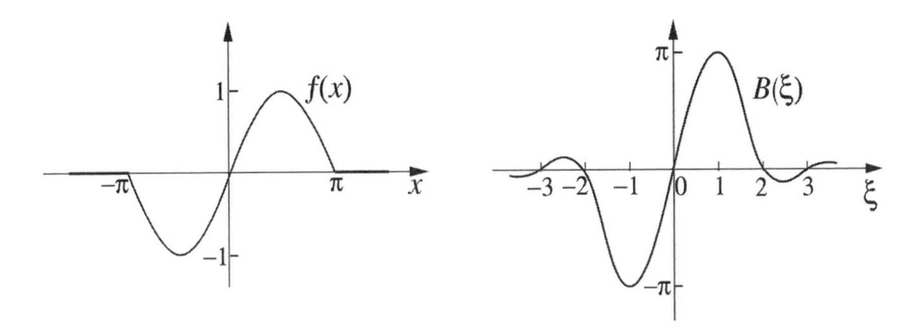

Figure 6.4 f and its sine transform.

By (6.28) the Fourier integral representation of f is therefore

$$f(x) = \frac{2}{\pi}\int_0^\infty \frac{\sin \pi\xi}{1-\xi^2}\sin x\xi \, d\xi. \tag{6.29}$$

Note here that

$$\lim_{\xi\to\pm 1}\frac{\sin \pi\xi}{1-\xi^2}\sin x\xi = \pm\frac{\pi}{2}\sin x,$$

so the integrand $(1-\xi^2)^{-1}\sin\pi\xi\sin x\xi$ is bounded on $0 \leq \xi < \infty$, and is dominated by c/ξ^2 as $\xi \to \infty$ for some positive constant c. The Fourier integral (6.29) therefore converges uniformly on $\mathbb{R}$ to the continuous function f.

Example 6.14

We have already seen in Example 6.3 that the Fourier transform of the even function $e^{-|x|}$ is

$$\mathcal{F}(e^{-|x|})(\xi) = \frac{2}{1+\xi^2},$$

which is an $\mathcal{L}^1$ function on $\mathbb{R}$, hence, using (6.26),

$$e^{-|x|} = \frac{2}{\pi}\int_0^\infty \frac{\cos x\xi}{1+\xi^2}d\xi.$$

At $x = 0$ we obtain the familiar result

$$\int_0^\infty \frac{d\xi}{1+\xi^2} = \frac{\pi}{2}.$$

We conclude this section with a brief look at the Fourier transformation in $\mathcal{L}^2$, the space in which the theory of Fourier series was developed in Chapter 3. Exercise 6.2 shows that, in general, there is no inclusion relation between

$\mathcal{L}^1(\mathbb{R})$ and $\mathcal{L}^2(\mathbb{R})$, but bounded functions in $\mathcal{L}^1(\mathbb{R})$ belong to $\mathcal{L}^2(\mathbb{R})$, because $\int |f|^2 \leq M \int |f|$ whenever $|f| \leq M$. Suppose that both f and g are piecewise smooth functions in $\mathcal{L}^1(\mathbb{R})$, and hence bounded, and that their transforms $\hat{f}$ and $\hat{g}$ also belong to $\mathcal{L}^1(\mathbb{R})$. According to Theorem 6.6, the functions $\hat{f}$ and $\hat{g}$ are continuous and bounded on $\mathbb{R}$. The improper integrals

$$\int_{-\infty}^{\infty} \hat{f}(\xi)e^{i\xi x}d\xi, \quad \int_{-\infty}^{\infty} \hat{g}(\xi)e^{i\xi x}d\xi$$

are also piecewise smooth on $\mathbb{R}$, for they represent the functions $2\pi f(x)$ and $2\pi g(x)$, respectively, by Theorem 6.10. Here, of course, we are assuming that f and g are defined at each point of discontinuity by the median of the "jump" at that point. Thus the functions f, g, $\hat{f}$, and $\hat{g}$ all belong to $\mathcal{L}^2(\mathbb{R})$, and we can write

$$\begin{aligned} 2\pi \langle f, g \rangle &= 2\pi \int_{-\infty}^{\infty} f(x)\overline{g(x)}dx \\ &= \int_{-\infty}^{\infty}\int_{-\infty}^{\infty} f(x)\overline{\hat{g}(\xi)e^{ix\xi}}d\xi dx \\ &= \int_{-\infty}^{\infty}\int_{-\infty}^{\infty} f(x)e^{-i\xi x}\overline{\hat{g}(\xi)}dx d\xi \\ &= \int_{-\infty}^{\infty} \hat{f}(\xi)\overline{\hat{g}(\xi)}d\xi \\ &= \left\langle \hat{f}, \hat{g} \right\rangle. \end{aligned} \tag{6.30}$$

When $g = f$, we obtain a relation between the $\mathcal{L}^2$ norms of f and its Fourier transform,

$$\left\|\hat{f}\right\|^2 = 2\pi \left\|f\right\|^2, \tag{6.31}$$

which corresponds to Parseval's relation (1.23). Equations (6.30) and (6.31) together constitute what is known as Plancherel's theorem, which actually holds under weaker conditions on f and g. In fact, this pair of equations is valid whenever f and g lie in $\mathcal{L}^2(\mathbb{R})$. The proof of this more general result is based on the fact that the set of functions $\mathcal{L}^2(\mathbb{R}) \cap \mathcal{L}^1(\mathbb{R})$ is dense in $\mathcal{L}^2(\mathbb{R})$, that is, that every function in $\mathcal{L}^2(\mathbb{R})$ is the limit (in $\mathcal{L}^2$) of a sequence of functions in $\mathcal{L}^2(\mathbb{R}) \cap \mathcal{L}^1(\mathbb{R})$.

EXERCISES

6.10 Express each of the following functions as a Fourier integral in trigonometric form:

(a) $f(x) = \begin{cases} |\sin x|, & |x| < \pi \\ 0, & |x| > \pi. \end{cases}$

(b) $f(x) = \begin{cases} 1, & 0 < x < 1 \\ 0, & x < 0,\ x > 1. \end{cases}$

(c) $f(x) = \begin{cases} 1 - x, & 0 < x < 1 \\ -1 - x, & -1 < x < 0 \\ 0, & |x| > 1. \end{cases}$

(d) $f(x) = \begin{cases} \cos x, & 0 < x < \pi/2 \\ -\cos x, & -\pi/2 < x < 0 \\ 0, & |x| > \pi/2. \end{cases}$

(e) $f(x) = \begin{cases} \cos x, & 0 < x < \pi \\ 0, & x < 0,\ x > \pi. \end{cases}$

6.11 Use the result of Exercise 6.10(e) to show that

$$\frac{\pi}{2} = \int_0^\infty \frac{\xi \sin \pi\xi}{1 - \xi^2} d\xi.$$

6.12 Prove that

$$e^{-\alpha x} = \frac{2}{\pi} \int_0^\infty \frac{\xi \sin \xi x}{\xi^2 + \alpha^2} d\xi \quad \text{for all} x > 0, \quad \alpha > 0.$$

Explain why the equality does not hold at $x = 0$.

6.13 Prove that

$$\int_0^\infty \frac{\xi^3 \sin x\xi}{\xi^4 + 4} d\xi = \frac{\pi}{2} e^{-x} \cos x \qquad \text{for all } x > 0.$$

Is the integral uniformly convergent on $[0, \infty)$?

6.14 Determine the Fourier cosine integral for $e^{-x} \cos x$, $x \geq 0$. Is the representation pointwise?

6.15 Prove that

$$\int_0^\infty \frac{1 - \cos \pi\xi}{\xi} \sin x\xi d\xi = \begin{cases} \pi/2, & 0 < x < \pi \\ \pi/4, & x = \pi \\ 0, & x > \pi \end{cases}$$

6.16 Determine the Fourier transform of $f(x) = (1 + x^2)^{-1}$, $x \in \mathbb{R}$.

6.17 Given

$$f(x) = \begin{cases} 1 - |x|, & |x| \leq 1 \\ 0, & |x| > 1, \end{cases}$$

show that

$$\hat{f}(\xi) = \left[\frac{\sin(\xi/2)}{\xi/2}\right]^2,$$

then conclude that

$$\int_{-\infty}^{\infty} \frac{\sin^2 \xi}{\xi^2} d\xi = \pi.$$

6.18 Verify the equality $\left\|\hat{f}\right\|^2 = 2\pi \|f\|^2$ in the case where $f(x) = e^{-|x|}$.

6.19 Express the relation (6.31) in terms of the cosine and sine transforms A and B.

6.3 Properties and Applications

The following theorem gives the fundamental properties of the Fourier transformation under differentiation. The formula for the transform of the derivative is particularly important, not only as a tool for solving linear partial differential equations, but also as a fundamental result on which the existence of solutions to such equations is based.

Theorem 6.15

Let $f \in \mathcal{L}^1(\mathbb{R})$.
(i) If $f' \in \mathcal{L}^1(\mathbb{R})$ and f is continuous on $\mathbb{R}$, then

$$\mathcal{F}(f')(\xi) = i\xi\mathcal{F}(f)(\xi), \qquad \xi \in \mathbb{R}. \tag{6.32}$$

(ii) If $xf(x) \in \mathcal{L}^1(\mathbb{R})$, then $\mathcal{F}(f)$ is differentiable and its derivative

$$\frac{d}{d\xi}\mathcal{F}(f)(\xi) = \mathcal{F}(-ixf)(\xi), \qquad \xi \in \mathbb{R}, \tag{6.33}$$

is continuous on $\mathbb{R}$.

Proof

(i) $|f'|$ being integrable, its Fourier transform $\mathcal{F}(f')$ exists and

$$\mathcal{F}(f')(\xi) = \int_{-\infty}^{\infty} f'(x)e^{-i\xi x}dx.$$

The continuity of f allows us to write $\int_0^x f'(t)dt = f(x) - f(0)$, hence the two limits

$$\lim_{x\to\pm\infty} f(x) = f(0) + \lim_{x\to\pm\infty} \int_0^x f'(t)dt$$

exist. But because f is continuous and integrable on $\mathbb{R}$, $\lim_{x\to\pm\infty} f(x) = 0$. Now integration by parts yields

$$\begin{aligned}\mathcal{F}(f')(\xi) &= f(x)e^{-i\xi x}\Big|_{-\infty}^{\infty} + i\xi \int_{-\infty}^{\infty} f(x)e^{-i\xi x}dx \\ &= i\xi\mathcal{F}(f)(\xi).\end{aligned}$$

(ii)

$$\frac{\hat{f}(\xi + \Delta\xi) - \hat{f}(\xi)}{\Delta\xi} = \int_{-\infty}^{\infty} f(x)\frac{e^{-i(\xi+\Delta\xi)x} - e^{-i\xi x}}{\Delta\xi}dx.$$

In the limit as $\Delta\xi \to 0$, we obtain

$$\frac{d}{d\xi}\mathcal{F}(f)(\xi) = \lim_{\Delta\xi\to 0} \int_{-\infty}^{\infty} f(x)\frac{e^{-i(\xi+\Delta\xi)x} - e^{-i\xi x}}{\Delta\xi}dx.$$

By hypothesis, the limit

$$\lim_{\Delta\xi\to 0} f(x)\frac{e^{-i(\xi+\Delta\xi)x} - e^{-i\xi x}}{\Delta\xi} = -ixf(x)e^{-i\xi x}$$

is dominated by $|xf(x)| \in \mathcal{L}^1(\mathbb{R})$ for every $\xi \in \mathbb{R}$, therefore we can use Theorem 6.4 to conclude that

$$\frac{d}{d\xi}\mathcal{F}(f)(\xi) = \int_{-\infty}^{\infty} [-ixf(x)]e^{-i\xi x}dx = \mathcal{F}(-ixf)(\xi).$$

□

Using induction, this result can easily be generalized.

Corollary 6.16

Suppose $f \in L^1(\mathbb{R})$ and n is any positive integer.

(i) If $f^{(k)} \in L^1(\mathbb{R})$ for all $1 \leq k \leq n$, and $f^{(n-1)}$ is continuous on $\mathbb{R}$, then

$$\mathcal{F}(f^{(n)})(\xi) = (i\xi)^n\mathcal{F}(f)(\xi). \tag{6.34}$$

(ii) If $x^n f(x) \in L^1(\mathbb{R})$, then

$$\frac{d^n}{d\xi^n}\mathcal{F}(f)(\xi) = \mathcal{F}[(-ix)^n f](\xi). \tag{6.35}$$

The integrability of $|x^n f(x)|$ on $\mathbb{R}$ may be viewed as a measure of how fast the function $f(x)$ tends to 0 as $x \to \pm\infty$, in the sense that $f(x)$ tends to 0 faster when n is larger. Equation (6.34) therefore implies that, as the order of differentiability (or smoothness) of f increases, so does the rate of decay of $\hat{f}$. Equation (6.35), on the other hand, indicates that functions which decay faster have smoother transforms.

Example 6.17

Let

$$f(x) = e^{-x^2}, \qquad x \in \mathbb{R}.$$

To determine the Fourier transform of f, we recall that

$$\int_{-\infty}^{\infty} e^{-x^2} dx = \sqrt{\pi} \tag{6.36}$$

(Exercise 4.15). Because f satisfies the conditions of Theorem 6.15, Equation (6.33) allows us to write

$$\begin{aligned}
\frac{d}{d\xi}\hat{f}(\xi) &= -i\int_{-\infty}^{\infty} xe^{-x^2}e^{-i\xi x}dx \\
&= \frac{i}{2}\int_{-\infty}^{\infty} \frac{d}{dx}\left(e^{-x^2}\right)e^{-i\xi x}dx \\
&= -\frac{i}{2}\int_{-\infty}^{\infty} e^{-x^2}(-i\xi)e^{-i\xi x}dx \\
&= -\frac{\xi}{2}\hat{f}(\xi).
\end{aligned}$$

The solution of this equation is

$$\hat{f}(\xi) = ce^{-\xi^2/4},$$

where the integration constant is determined by setting $\xi = 0$ and using Equation (6.36); that is,

$$c = \hat{f}(0) = \sqrt{\pi}.$$

Thus $\mathcal{F}(e^{-x^2})(\xi) = \sqrt{\pi}e^{-\xi^2/4}$.

6.3.1 Heat Transfer in an Infinite Bar

Just as Fourier series served us in the construction of solutions to boundary-value problems in bounded space domains, we now show how such solutions can be represented by Fourier integrals when the space domain becomes unbounded. The question of the uniqueness of a solution obtained in this manner, in general, is not addressed here, as it properly belongs to the theory of partial differential equations. But the equations that we have already introduced (Laplace's equation, the heat equation, and the wave equation) all have unique solutions under the boundary conditions imposed, whether the space variable is bounded or not. This follows from the fact that, due to the linearity of these equations and that of their boundary conditions, the difference between any two solutions of a problem satisfies a homogeneous differential equation under homogeneous boundary conditions, which can only have a trivial solution (see [13]).

Example 6.18

Suppose that an infinite thin bar has an initial temperature distribution along its length given by $f(x)$. We wish to determine the temperature $u(x,t)$ along the bar for all $t>0$. To solve this problem by the Fourier transform, we assume that f is piecewise smooth and that $|f|$ is integrable on $(-\infty,\infty)$.

The temperature $u(x,t)$ satisfies the heat equation

$$u_t = ku_{xx}, \quad -\infty < x < \infty, \quad t>0, \tag{6.37}$$

and the initial condition

$$u(x,0)=f(x), \quad -\infty < x < \infty. \tag{6.38}$$

As before, we resort to separation of variables by assuming that

$$u(x,t)=v(x)w(t).$$

Substituting into Equation (6.37), we obtain the equation

$$\frac{v''(x)}{v(x)} = \frac{1}{k}\frac{w'(t)}{w(t)}, \quad -\infty < x < \infty, \quad t>0,$$

which implies that each side must be a constant, say $-\lambda^2$. The resulting pair of equations leads to the solutions

$$v(x) = A(\lambda)\cos\lambda x + B(\lambda)\sin\lambda x,$$
$$w(t) = C(\lambda)e^{-k\lambda^2 t},$$

where A, B, and C are constants of integration which depend on the parameter λ. Since there are no boundary conditions on the solution, λ will be a real (rather than a discrete) variable, and $A=A(\lambda)$ and $B=B(\lambda)$ will therefore be functions of λ, and we can set $C(\lambda)=1$ in the product $v(x)w(t)$. Corresponding to each $\lambda\in\mathbb{R}$, the function

$$u_\lambda(x,t) = [A(\lambda)\cos\lambda x + B(\lambda)\sin\lambda x]e^{-k\lambda^2 t}$$

therefore satisfies Equation (6.37), $A(\lambda)$ and $B(\lambda)$ being arbitrary functions of $\lambda \in \mathbb{R}$. We do not lose any generality by assuming that $\lambda \geq 0$ because the negative values of λ do not generate additional solutions.

$u_\lambda(x,t)$ cannot be expected to satisfy the initial condition (6.38) for a general function f, so we assume that the desired solution has the form

$$\begin{aligned} u(x,t) &= \frac{1}{\pi}\int_0^\infty u_\lambda(x,t)d\lambda \\ &= \frac{1}{\pi}\int_0^\infty [A(\lambda)\cos\lambda x + B(\lambda)\sin\lambda x]e^{-k\lambda^2 t}d\lambda. \end{aligned} \tag{6.39}$$

This assumption is suggested by the summation (3.30) over the discrete values of λ, and the factor $1/\pi$ is introduced in order to express u in the form of a Fourier integral. At $t = 0$, we have

$$\begin{aligned} u(x,0) &= \frac{1}{\pi}\int_0^\infty [A(\lambda)\cos\lambda x + B(\lambda)\sin\lambda x]d\lambda \\ &= f(x), \quad x \in \mathbb{R}. \end{aligned} \tag{6.40}$$

By Theorem 6.10 and Equations (6.22) to (6.24), we see that Equation (6.40) uniquely determines A and B as the cosine and sine transforms, respectively, of f :

$$\begin{aligned} A(\lambda) &= \int_{-\infty}^\infty f(y)\cos\lambda y \, dy \\ B(\lambda) &= \int_{-\infty}^\infty f(y)\sin\lambda y \, dy. \end{aligned}$$

Substituting back into (6.39), and using the even and odd properties of A and B, we arrive at the solution of the heat equation (6.37) which satisfies the initial condition (6.38)

$$\begin{aligned} u(x,t) &= \frac{1}{2\pi}\int_{-\infty}^\infty\int_{-\infty}^\infty f(y)[\cos\lambda y\cos\lambda x + \sin\lambda y\sin\lambda x]e^{-k\lambda^2 t}dyd\lambda \\ &= \frac{1}{\pi}\int_0^\infty\int_{-\infty}^\infty f(y)\cos(x-y)\lambda \, e^{-k\lambda^2 t}dyd\lambda. \end{aligned} \tag{6.41}$$

The solution (6.41) can also be obtained by first taking the Fourier transform of both sides of the heat equation, as functions of x, and using Corollary 6.16 to obtain

$$\hat{u}_t = k(i\xi)^2\hat{u}(\xi,t) = -k\xi^2\hat{u}(\xi,t).$$

The solution of this equation is

$$\hat{u}(\xi,t) = ce^{-k\xi^2 t},$$

where, by the initial condition, $c = \hat{u}(\xi, 0) = \hat{f}(\xi)$. Thus, as a function in ξ,

$$\hat{u}(\xi, t) = \hat{f}(\xi)e^{-k\xi^2 t}$$

is the Fourier transform of $u(x,t)$ for every $t > 0$, and u can therefore be represented, according to Theorem 6.10, by the integral

$$\begin{aligned}
u(x,t) &= \frac{1}{2\pi}\int_{-\infty}^{\infty} \hat{f}(\xi)e^{-k\xi^2 t}e^{i\xi x}d\xi \\
&= \frac{1}{2\pi}\int_{-\infty}^{\infty}\left[\int_{-\infty}^{\infty} f(y)e^{-i\xi y}dy\right]e^{-k\xi^2 t}e^{i\xi x}d\xi \\
&= \frac{1}{2\pi}\int_{-\infty}^{\infty}\int_{-\infty}^{\infty} f(y)e^{i(x-y)\xi}e^{-k\xi^2 t}d\xi dy.
\end{aligned}$$

Here the fact that $\left|\hat{f}(\xi)\right| e^{-k\xi^2 t}$ is integrable on $-\infty < \xi < \infty$ allows us to replace the Cauchy principal value in (6.16) by the corresponding improper integral, and the change of the order of integration in the last step is justified by the assumption that $|f|$ is integrable. Now, because $e^{-k\xi^2 t}$ is an even function of ξ, we have

$$u(x,t) = \frac{1}{\pi}\int_{-\infty}^{\infty} f(y)\int_{0}^{\infty} \cos(x-y)\xi \; e^{-k\xi^2 t}d\xi dy,$$

which coincides with (6.41).

Using the integration formula (see Exercise 6.23)

$$\int_0^\infty e^{-b\xi^2}\cos z\xi \; d\xi = \frac{1}{2}\sqrt{\frac{\pi}{b}}e^{-z^2/4b} \quad \text{for all } z \in \mathbb{R}, \quad b > 0, \tag{6.42}$$

we therefore have

$$\begin{aligned}
u(x,t) &= \frac{1}{2\sqrt{\pi k t}}\int_{-\infty}^{\infty} f(y)e^{-(y-x)^2/4kt}dy \\
&= \frac{1}{\sqrt{\pi}}\int_{-\infty}^{\infty} f\left(x + 2\sqrt{kt}p\right)e^{-p^2}dp.
\end{aligned}$$

Integrating by parts, it is straightforward to verify that this last expression for $u(x,t)$ satisfies the heat equation. It also satisfies the initial condition $u(x,0) = f(x)$ by the result of Exercise 4.15.

Example 6.19

The corresponding boundary-value problem for a semi-infinite bar, which is insulated at one end, is defined by the system of equations

$$u_t = ku_{xx}, \quad 0 < x < \infty, \quad t > 0 \tag{6.43}$$

$$u_x(0,t) = 0, \quad t > 0 \tag{6.44}$$

$$u(x,0) = f(x), \quad 0 < x < \infty. \tag{6.45}$$

The solution of Equation (6.43) by separation of variables leads to the set of solutions obtained in Example 6.18,

$$u_\lambda(x,t) = [A(\lambda)\cos \lambda x + B(\lambda)\sin \lambda x]e^{-k\lambda^2 t}, \quad 0 \le \lambda < \infty.$$

For each solution in this set to satisfy the boundary condition at $x = 0$, we must have

$$\left.\frac{\partial u_\lambda}{\partial x}\right|_{x=0} = \lambda B(\lambda)e^{-k\lambda^2 t} = 0 \quad \text{for all } t > 0.$$

If $\lambda = 0$ we obtain the constant solution $u_0 = A(0)$, and if $B(\lambda) = 0$ the solution is given by

$$u_\lambda(x,t) = A(\lambda)\cos \lambda x \; e^{-k\lambda^2 t}. \tag{6.46}$$

That means (6.46), where $0 \le \lambda < \infty$, gives all the solutions of the heat equation which satisfy the boundary condition (6.44).

In order to satisfy the initial condition (6.45), we form the integral

$$u(x,t) = \frac{1}{\pi}\int_0^\infty u_\lambda(x,t)d\lambda = \frac{1}{\pi}\int_0^\infty A(\lambda)\cos x\lambda \; e^{-k\lambda^2 t}d\lambda.$$

When $t = 0$,

$$u(x,0) = f(x) = \frac{1}{\pi}\int_0^\infty A(\lambda)\cos x\lambda \; d\lambda. \tag{6.47}$$

By extending f as an even function into $(-\infty,\infty)$, we see from the representation (6.47) that $A(\lambda)$ is the cosine transform of f. Hence, with $f \in \mathcal{L}^1(\mathbb{R})$,

$$A(\lambda) = \int_{-\infty}^\infty f(y)\cos \lambda y \; dy = 2\int_0^\infty f(y)\cos \lambda y \; dy,$$

and the solution of the boundary-value problem is given by

$$u(x,t) = \frac{2}{\pi}\int_0^\infty \int_0^\infty f(y)\cos \lambda y \cos \lambda x \; e^{-k\lambda^2 t}dyd\lambda. \tag{6.48}$$

Using the identity $2\cos \lambda y \cos \lambda x = \cos \lambda(y-x) + \cos \lambda(y+x)$ and the formula (6.42), this double integral may be reduced to the single integral representation

$$u(x,t) = \frac{1}{2\sqrt{\pi kt}}\int_0^\infty f(y)\left[e^{-(y-x)^2/4kt} + e^{-(y+x)^2/4kt}\right]dy.$$

To obtain an explicit expression for the solution when

$$f(y) = \begin{cases} 1, & 0 < y < a \\ 0, & y > a, \end{cases}$$

we can use the definition of the error function given in Exercise 5.6,

$$\text{erf}(x) = \frac{2}{\sqrt{\pi}} \int_0^x e^{-p^2} dp,$$

to write

$$\begin{aligned} u(x,t) &= \frac{1}{2\sqrt{\pi kt}} \int_0^a \left[e^{-(y-x)^2/4kt} + e^{-(y+x)^2/4kt} \right] dy \\ &= \frac{1}{\sqrt{\pi}} \int_{-x/2\sqrt{kt}}^{(a-x)/2\sqrt{kt}} e^{-p^2} dp + \frac{1}{\sqrt{\pi}} \int_{x/2\sqrt{kt}}^{(a+x)/2\sqrt{kt}} e^{-p^2} dp \\ &= \frac{1}{\sqrt{\pi}} \int_0^{(a-x)/2\sqrt{kt}} e^{-p^2} dp + \frac{1}{\sqrt{\pi}} \int_0^{(a+x)/2\sqrt{kt}} e^{-p^2} dp \\ &= \frac{1}{2} \text{erf} \left(\frac{a-x}{2\sqrt{kt}} \right) + \frac{1}{2} \text{erf} \left(\frac{a+x}{2\sqrt{kt}} \right). \end{aligned}$$

From the properties of the error function (Exercise 5.6) it is straightforward to verify that, for all $t > 0$,

$$u(x,t) \to 0 \quad \text{as } x \to \infty,$$

$$u(x,t) \to \text{erf}\left(a/2\sqrt{kt}\right) \quad \text{as} \quad x \to 0,$$

and that, for all $x > 0$,

$$u(x,t) \to 0 \quad \text{as} \quad t \to \infty.$$

As $t \to 0$,

$$u(x,t) \to \begin{cases} 1 & \text{when } 0 \le x < a, \\ 0 & \text{when } x > a. \end{cases}$$

Thus, at any point on the bar, the temperature approaches 0 as $t \to \infty$; and at any instant, the temperature approaches 0 as $x \to \infty$. This is to be expected, because the initial heat in the bar eventually seeps out to ∞. It is also worth noting that $u(a,t) \to 1/2 = [f(a^+) + f(a^-)]/2$ as $t \to 0$, as would be expected (see Figure 6.5).

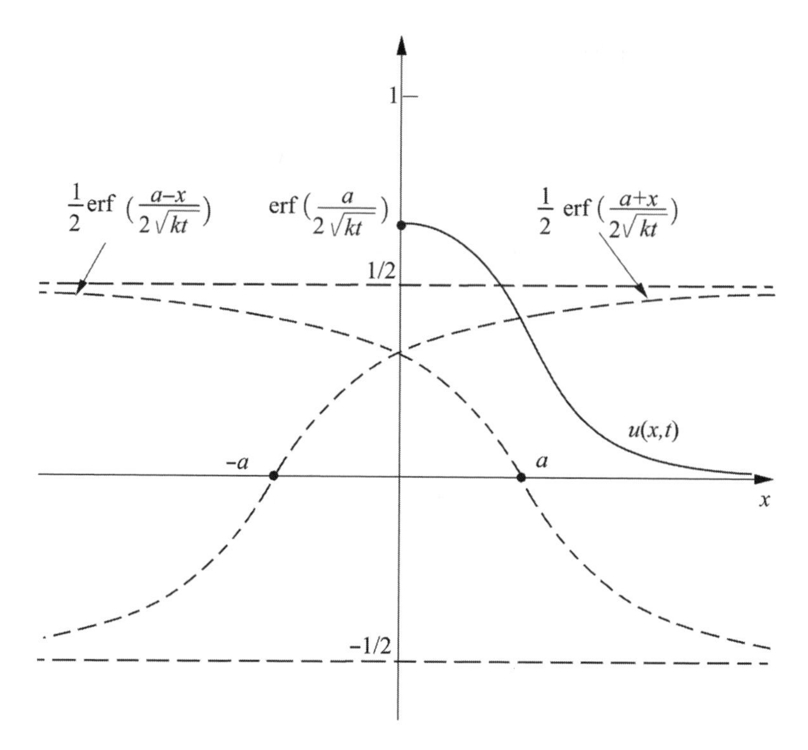

Figure 6.5 Temperature distribution on a semi-infinite rod.

6.3.2 Non-Homogeneous Equations

The nonhomogeneous differential equation $y - y'' = f$ can be solved directly by well-known methods when f is a polynomial or an exponential function, for example, but such methods do not work for more general classes of functions. If f has a Fourier transform, we can use Corollary 6.16 to write

$$\hat{y} + \xi^2 \hat{y} = \hat{f}.$$

Thus the solution of the differential equation is given, formally, by

$$y(x) = \mathcal{F}^{-1}\left(\hat{f}(\xi)\frac{1}{\xi^2+1}\right)(x). \tag{6.49}$$

We know the inverse transforms of both $\hat{f}$ and $(\xi^2+1)^{-1}$, but we have no obvious method for inverting the product of these two transforms, or even deciding whether such a product is invertible. Now we show how we can express $\mathcal{F}^{-1}(\hat{f}\cdot\hat{g})$ in terms of f and g under relatively mild restrictions on the functions f and g.

Definition 6.20

Let I be a real interval. A function $f : I \to \mathbb{C}$ is said to be *locally integrable* on I if $|f|$ is integrable on any finite subinterval of I.

Thus all piecewise continuous functions on I and all functions in $\mathcal{L}^1(I)$ are locally integrable.

Definition 6.21

If the functions $f, g : \mathbb{R} \to \mathbb{C}$ are locally integrable, their *convolution* is the function defined by the integral

$$f * g(x) = \int_{-\infty}^{\infty} f(x-t)g(t)dt$$

for all $x \in \mathbb{R}$ where the integral converges.

By setting $x - t = s$ in the above integral we obtain the commutative relation

$$f * g(x) = \int_{-\infty}^{\infty} f(s)g(x-s)ds = g * f(x).$$

The convolution $f * g$ exists as a function on $\mathbb{R}$ under various conditions on f and g. Here are some examples.
1. If either function is absolutely integrable and the other is bounded. Suppose $f \in \mathcal{L}^1(\mathbb{R})$ and $|g| \leq M$. Then

$$\begin{aligned} |f * g(x)| &\leq \int_{-\infty}^{\infty} |f(x-t)|\,|g(t)|\,dt \\ &\leq M \int_{-\infty}^{\infty} |f(t)|\,dt < \infty. \end{aligned}$$

2. If both f and g vanish on $(-\infty, 0)$, in which case

$$f * g(x) = \int_{-\infty}^{\infty} f(x-t)g(t)dt = \int_0^x f(x-t)g(t)dt.$$

3. If both f and g belong to $\mathcal{L}^2(\mathbb{R})$. This follows directly from the CBS inequality.
4. If either f or g is bounded and vanishes outside a finite interval, then $f * g$ is clearly bounded.

But we need more than the boundedness of $f * g$ in order to proceed.

Theorem 6.22

If both f and g belong to $\mathcal{L}^1(\mathbb{R})$ and either function is bounded, then $f * g$ also lies in $\mathcal{L}^1(\mathbb{R})$ and

$$\mathcal{F}(f * g) = \hat{f}\hat{g}. \tag{6.50}$$

Proof

For $f * g$ to have a Fourier transform we first have to prove that $|f * g|$ is integrable on $\mathbb{R}$. Suppose, without loss of generality, that $|g|$ is bounded on $\mathbb{R}$ by the positive constant M. Because

$$|f(x-t)g(t)| \leq M\,|f(x-t)|$$

and $M\,|f(x-t)|$ is integrable (as a function of x) on $(-\infty, \infty)$ for all $t \in \mathbb{R}$, it follows from Lemma 6.9 that the integral

$$f * g(x) = \int_{-\infty}^{\infty} f(x-t)g(t)dt$$

is uniformly convergent. Consequently, for any $a > 0$, we can write

$$\begin{aligned}
\int_0^a |f * g(x)|\, dx &= \int_0^a \left| \int_{-\infty}^{\infty} f(x-t)g(t)dt \right| dx \\
&\leq \int_0^a \int_{-\infty}^{\infty} |f(x-t)g(t)|\, dtdx \\
&= \int_{-\infty}^{\infty} \int_0^a |f(x-t)g(t)|\, dxdt \\
&= \int_{-\infty}^{\infty} |g(t)| \int_0^a |f(x-t)|\, dxdt \\
&\leq \int_{-\infty}^{\infty} |g(t)| \int_{-\infty}^{\infty} |f(x-t)|\, dxdt \\
&= \int_{-\infty}^{\infty} |g(t)|\, dt \int_{-\infty}^{\infty} |f(x)|\, dx.
\end{aligned}$$

This inequality holds for any $a > 0$, therefore $|f * g|$ is integrable on $(0, \infty)$. Similarly, $|f * g|$ is integrable on $(-\infty, 0)$ and hence on $\mathbb{R}$.

To prove the equality (6.50) we write

$$\begin{aligned}
\mathcal{F}(f*g)(\xi) &= \int_{-\infty}^{\infty}\int_{-\infty}^{\infty} f(x-t)g(t)e^{-i\xi x}dtdx \\
&= \int_{-\infty}^{\infty}\int_{-\infty}^{\infty} f(x-t)e^{-i\xi(x-t)}g(t)e^{-i\xi t}dxdt \\
&= \int_{-\infty}^{\infty} f(s)e^{-i\xi s}ds \int_{-\infty}^{\infty} g(t)e^{-i\xi t}dt \\
&= \hat{f}(\xi)\hat{g}(\xi),
\end{aligned}$$

where the change in the order of integration is justified by the uniform convergence of the convolution integral. □

Using Equation (6.50) and the result of Example 6.3 we therefore conclude that a particular solution of $y - y'' = f$ is given by

$$\begin{aligned}
y(x) &= \mathcal{F}^{-1}\left(\hat{f}(\xi)\frac{1}{\xi^2+1}\right)(x) \\
&= f*\left(\frac{1}{2}e^{-|\cdot|}\right)(x) && (6.51) \\
&= \frac{1}{2}\int_{-\infty}^{\infty} f(x-t)e^{-|t|}dt \\
&= \frac{1}{2}\int_{-\infty}^{\infty} f(t)e^{-|x-t|}dt. && (6.52)
\end{aligned}$$

We leave it as an exercise to verify, by direct differentiation, that this integral expression satisfies the equation $y - y'' = f$, provided f is continuous.

In this connection, it is worth noting that the kernel function $e^{-|x-t|}$ in the integral representation of y in Equation (6.52) is no other than Green's function $G(x,t)$ for the SL operator

$$L = -\frac{d^2}{dx^2} + 1 \qquad (6.53)$$

on the interval $(-\infty,\infty)$, subject to the boundary conditions $\lim_{x\to\pm\infty} G(x,t) = 0$; for the nonhomogeneous equation $Ly = f$ is solved, formally, by

$$y(x) = L^{-1}f = \int_{-\infty}^{\infty} G(x,t)f(t)dt.$$

This solution tends to 0 as $x \to \pm\infty$, as it should, being continuous and integrable on $(-\infty,\infty)$. Under such boundary conditions, the solution of the corresponding homogeneous equation $y - y'' = 0$, namely $c_1e^x + c_2e^{-x}$, can only be the trivial solution.

To solve the same equation $y - y'' = f$ on the semi-infinite interval $[0, \infty)$, we need to impose a boundary condition at $x = 0$, say

$$y(0) = y_0.$$

In this case the same procedure above leads to the particular solution

$$y_p(x) = \frac{1}{2}\int_0^\infty f(t)e^{-|x-t|}dt,$$

but here the homogeneous solution

$$y_h(x) = ce^{-x}, \quad x \geq 0$$

is admissible for any constant c. Applying the boundary condition at $x = 0$ to the sum $y = y_p + y_h$ yields

$$y_0 = \frac{1}{2}\int_0^\infty f(t)e^{-t}dt + c,$$

from which c can be determined. The desired solution is therefore

$$y(x) = \frac{1}{2}\int_0^\infty f(t)e^{-|x-t|}dt + \left(y_0 - \frac{1}{2}\int_0^\infty f(t)e^{-t}dt\right)e^{-x}.$$

If $y_0 = 0$, that is, if the boundary condition at $x = 0$ is homogeneous, then

$$y(x) = \frac{1}{2}\int_0^\infty f(t)\left(e^{-|x-t|} - e^{-(x+t)}\right)dt,$$

and, once again, we conclude that Green's function for the same operator (6.53) on the interval $[0, \infty)$ under the homogeneous boundary condition $y(0) = 0$ is now

$$G(x,t) = \frac{1}{2}(e^{-|x-t|} - e^{-(x+t)}).$$

In Chapter 7 we show that the Laplace transformation provides a more effective tool for solving nonhomogeneous differential equations in $[0, \infty)$, especially where the coefficients are constants.

EXERCISES

6.20 Prove the following formulas.

$$\mathcal{F}[f(x-a)](\xi) = e^{-ia\xi}\hat{f}(\xi)$$
$$\mathcal{F}[e^{iax}f(x)](\xi) = \hat{f}(\xi - a).$$

6.21 The *Hermite function* of order n, for any $n \in \mathbb{N}_0$, is defined by

$$\psi_n(x) = e^{-x^2/2} H_n(x), \quad -\infty < x < \infty,$$

where H_n is the Hermite polynomial of order n. Prove that $\hat{\psi}_n$ exists and satisfies

$$\hat{\psi}_n(\xi) = (-i)^n \sqrt{2\pi}\psi_n(\xi).$$

Hint: Use Example 6.17 and induction on n

6.22 Solve the integral equation

$$\int_0^\infty u(x)\cos\xi x dx = \begin{cases} 1, & 0 < \xi < \pi \\ 0, & \pi < \xi < \infty. \end{cases}$$

6.23 Prove the integration formula (6.42). Hint: Denote the integral by $I(z)$, show that it satisfies the differential equation $2bI'(z) + zI(z) = 0$, and then solve this equation.

6.24 Assuming that

$$f(x) = \begin{cases} T_0, & |x| < a \\ 0, & |x| > a \end{cases}$$

in Example 6.18, prove that

$$u(x,t) = \frac{T_0}{2}\left[\operatorname{erf}\left(\frac{x+a}{2\sqrt{kt}}\right) + \operatorname{erf}\left(\frac{a-x}{2\sqrt{kt}}\right)\right].$$

Compare this to the solution in Example 6.19, and explain the similarity between the two.

6.25 Show that the solution of the boundary-value problem

$$\begin{aligned} u_t &= ku_{xx}, && 0 < x < \infty,\ t > 0 \\ u(0,t) &= 0, && t > 0 \\ u(x,0) &= f(x), && 0 < x < \infty, \end{aligned}$$

where f is a piecewise smooth function in $\mathcal{L}^1(0,\infty)$, is given by

$$u(x,t) = \frac{1}{2\sqrt{\pi kt}} \int_0^\infty f(y)\left[e^{-(y-x)^2/4kt} - e^{-(y+x)^2/4kt}\right] dy.$$

This is a mathematical model for heat flow in a semi-infinite rod with one end held at zero temperature.

6.26 If $f(x) = 1$ on $(0,a)$ and 0 otherwise in Exercise 6.25, express $u(x,t)$ in terms of the error function.

6.27 Use the Fourier transform to solve the initial-value problem for the wave equation

$$\begin{aligned} u_{tt} &= c^2 y_{xx}, & -\infty < x < \infty, \quad t > 0 \\ u(x,0) &= f(x), & -\infty < x < \infty \\ u_t(x,0) &= 0, & -\infty < x < \infty, \end{aligned}$$

where f is assumed to be a piecewise smooth function in $\mathcal{L}^1(\mathbb{R})$. Show that the solution can be expressed as

$$u(x,t) = \frac{1}{2\pi} \int_{-\infty}^{\infty} \hat{f}(\xi) \cos ct\xi \; e^{ix\xi} d\xi,$$

and derive the d'Alembert representation

$$u(x,t) = \frac{1}{2}[f(x+ct) + f(x-ct)].$$

6.28 Solve the wave equation on the semi-infinite domain $0 < x < \infty$ under the initial conditions in Exercise 6.27 and the boundary condition $u(0,t) = 0$ for all $t > 0$.

6.29 Verify, by direct differentiation, that the integral expression (6.52) solves the equation $y - y'' = f$.

7 The Laplace Transformation

If we replace the imaginary variable $i\xi$ in the Fourier transform

$$\hat{f}(\xi) = \int_{-\infty}^{\infty} f(x) e^{-i\xi x} dx$$

by the complex variable $s = \sigma + i\xi$, and set $f(x) = 0$ for all $x < 0$, the function defined by the resulting integral,

$$F(s) = \int_{0}^{\infty} f(x) e^{-sx} dx, \qquad (7.1)$$

is called the Laplace transform of f. When $\operatorname{Re} s = \sigma$ is positive, the improper integral in Equation (7.1) exists even if $|f|$ is not integrable on $(0, \infty)$, such as when f is a polynomial, and herein lies the advantage of the Laplace transform. Because of the exponential decay of $e^{-\sigma x}$, for $\sigma > 0$, the Laplace transformation is defined on a much larger class of functions than $\mathcal{L}^1(0, \infty)$.

7.1 The Laplace Transform

Definition 7.1

We use $\mathcal{E}$ to denote the class of functions $f : [0, \infty) \to \mathbb{C}$ such that f is locally integrable in $[0, \infty)$ and $f(x)e^{-\alpha x}$ remains bounded as $x \to \infty$ for some real number α.

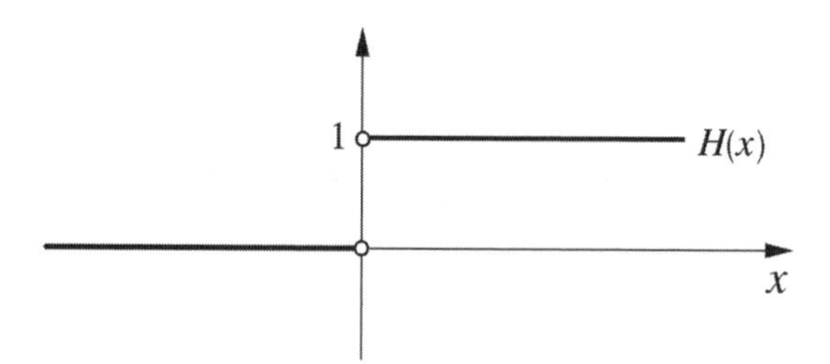

Figure 7.1 The Heaviside function.

Thus a function $f \in \mathcal{E}$ is characterized by two properties: first, that f is allowed to have singular points in $[0, \infty)$ where f is nevertheless locally integrable; and, second, that f has at most exponential growth as $x \to \infty$, that is, that there is a real number α such that $f(x)$ is dominated by a constant multiple of the exponential function $e^{-\alpha x}$ for large values of x. More precisely, there are positive constants b and c such that

$$|f(x)| \leq ce^{\alpha x} \qquad \text{for all } x \geq b. \tag{7.2}$$

For the sake of convenience, $\mathcal{E}$ will also include any function defined on $\mathbb{R}$ which vanishes on $(-\infty, 0)$ and satisfies Definition 7.1 on $[0, \infty)$. If we now define the Heaviside function H on $\mathbb{R}\backslash\{0\}$ by (see Figure 7.1)

$$H(x) = \begin{cases} 1, & x > 0 \\ 0, & x < 0, \end{cases}$$

then the following functions all lie in $\mathcal{E}$.

(i) $H(x)g(x)$ for any bounded locally integrable function g on $\mathbb{R}$, such as $\cos x$ or $\sin x$. Here α, as referred to in Definition 7.1, can be any nonnegative number.

(ii) $H(x)p(x)$ for any polynomial p. Here α is any positive number.

(iii) $H(x)e^{kx}$, $\alpha \geq k$.

(iv) $H(x)\log x$, α any positive number.

(v) $H(x)x^{\mu}$, $\mu > -1$, α any positive number.

But such functions as $H(x)e^{x^2}$ or $H(x)2^{3^x}$ do not belong to $\mathcal{E}$. In the above examples of functions in $\mathcal{E}$, we take $f(0)$ to be $f(0^+)$, and it is of no consequence how f is defined at $x = 0$. That is why we chose not to define H at that point. The rate of growth of a function f is characterized by the smallest value of α which makes $f(x)e^{-\alpha x}$ bounded as $x \to \infty$. When such a smallest value exists, it is referred to as the exponential order of f, such as $\alpha = 0$ in (i) and $\alpha = k$ in (iii). In (ii), (iv), and (v) α can be any positive number. A function which satisfies the inequality (7.2) is said to be of exponential order.

Definition 7.2

For any $f \in \mathcal{E}$, the *Laplace transform* of f is defined for all $\operatorname{Re} s = \sigma > \alpha$ by the improper integral

$$\mathcal{L}(f)(s) = \int_0^\infty f(x)e^{-sx}dx. \tag{7.3}$$

Basically, this definition is tailored to guarantee that $f(x)e^{-sx}$ lies in $\mathcal{L}^1(0,\infty)$. By the estimate (7.2),

$$\int_0^\infty \left| f(x)e^{-(\sigma+i\xi)x} \right| dx \leq c\int_0^\infty e^{-(\sigma-\alpha)x}dx < \infty \qquad \text{for all } \sigma > \alpha,$$

we see that the integral (7.3) converges for all $\sigma > \alpha$. The convergence is uniform in the half-plane $\sigma \geq \alpha + \varepsilon$ for any $\varepsilon > 0$, where the integral can be differentiated with respect to s by passing under the integral sign. The resulting integral

$$-\int_0^\infty xf(x)e^{-sx}dx$$

also converges in $\sigma \geq \alpha + \varepsilon$. Consequently the Laplace transform $\mathcal{L}(f)$ is an analytic function of the complex variable s in the half-plane $\operatorname{Re} s > \alpha$. As in the case of the Fourier transformation, the Laplace transformation $\mathcal{L}$ is linear on the class of functions $\mathcal{E}$, so that

$$\mathcal{L}(af + bg) = a\mathcal{L}(f) + b\mathcal{L}(g), \qquad a, b \in \mathbb{C}, \qquad f, g \in \mathcal{E}.$$

Example 7.3

The constant function $f(x) = 1$ on $[0,\infty)$ clearly belongs to $\mathcal{E}$, with exponential order $\alpha = 0$, and its Laplace transform is

$$\mathcal{L}(1)(s) = \int_0^\infty e^{-sx}dx = \frac{1}{s}, \qquad s \in \mathbb{C}, \qquad \operatorname{Re} s > 0.$$

Similarly, the function $f : [0,\infty) \to \mathbb{R}$ defined by $f(x) = x^n$, $n \in \mathbb{N}$, also belongs to $\mathcal{E}$, with $\alpha > 0$. For all $\operatorname{Re} s > 0$ we have, using integration by parts,

$$\begin{aligned}\mathcal{L}(x^n)(s) &= \int_0^\infty x^n e^{-sx}dx \\ &= \frac{n}{s}\int_0^\infty x^{n-1}e^{-sx}dx \\ &= \cdots\end{aligned}$$

$$
\begin{aligned}
&= \frac{n!}{s^n}\int_0^\infty e^{-sx}dx \\
&= \frac{n!}{s^{n+1}}. \\
\Rightarrow \quad \mathcal{L}(x^n)(s) &= \frac{n!}{s^{n+1}}, \qquad n \in \mathbb{N}_0.
\end{aligned}
$$

The integrability of $|f(x)e^{-sx}|$ does not depend on $\operatorname{Im} s = \xi$, therefore it is more convenient, for the sake of evaluating the integral (7.3), to set $\operatorname{Im} s = 0$. Being analytic on $\operatorname{Re} s > \alpha$ in the complex s-plane, the function $F = \mathcal{L}(f)$ is uniquely determined by its values on the real axis $s > \alpha$.

Example 7.4

For any $\mu > -1$, the Laplace transform

$$
\mathcal{L}(x^\mu) = \int_0^\infty x^\mu e^{-sx}dx
$$

exists because the singularity of x^μ as $x \to 0^+$ is integrable. To evaluate this indefinite integral, we take s to be real and set $sx = t$. If s is positive,

$$
\begin{aligned}
\mathcal{L}(x^\mu) &= \int_0^\infty \left(\frac{t}{s}\right)^\mu e^{-t}\frac{dt}{s} \\
&= \frac{1}{s^{\mu+1}}\int_0^\infty e^{-t}t^\mu dt \\
&= \frac{1}{s^{\mu+1}}\Gamma(\mu+1), \qquad s > 0.
\end{aligned}
$$

This result can now be extended analytically into $\operatorname{Re} s > 0$, and it clearly generalizes that of Example 7.3.

Example 7.5

For any real number a,

$$
\mathcal{L}(e^{ax})(s) = \int_0^\infty e^{-(s-a)x}dx = \frac{1}{s-a}, \qquad s > a.
$$

Hence,

$$
\begin{aligned}
\mathcal{L}(\sinh ax)(s) &= \mathcal{L}\left(\frac{1}{2}e^{ax} - \frac{1}{2}e^{-ax}\right) \\
&= \frac{1}{2}\left(\frac{1}{s-a} - \frac{1}{s+a}\right) \\
&= \frac{a}{s^2-a^2}, \quad s > |a|.
\end{aligned}
$$

On the other hand,

$$\mathcal{L}(e^{iax})(s) = \int_0^\infty e^{-(s-ia)x}dx = \frac{1}{s-ia}, \qquad s > 0.$$

Therefore

$$\begin{aligned}\mathcal{L}(\sin ax)(s) &= \frac{1}{2i}\left(\frac{1}{s-ia} - \frac{1}{s+ia}\right) \\ &= \frac{a}{s^2+a^2}, \qquad s > 0.\end{aligned}$$

To recover a function f from its Laplace transform $\mathcal{L}(f) = F$, we resort to the Fourier inversion formula (6.16). Suppose $f \in \mathcal{E}$ so that $|f(x)e^{-\alpha x}|$ is integrable on $[0,\infty)$ for some real number α. Choose $\beta > \alpha$ and define the function

$$g(x) = \begin{cases} f(x)e^{-\beta x}, & x \geq 0 \\ 0, & x < 0, \end{cases}$$

which clearly belongs to $\mathcal{L}^1(\mathbb{R})$. Its Fourier transform is given by

$$\hat{g}(\xi) = \int_0^\infty f(x)e^{-\beta x}e^{-i\xi x}dx = \mathcal{L}(f)(\beta + i\xi) = F(\beta + i\xi).$$

If we further assume that f is piecewise smooth on $[0,\infty)$, then g will also be piecewise smooth on $\mathbb{R}$, and we can define g to be the average of its left and right limits at each point of discontinuity. By Theorem 6.9,

$$\begin{aligned}g(x) &= f(x)e^{-\beta x} \\ &= \lim_{L\to\infty} \frac{1}{2\pi}\int_{-L}^{L} F(\beta + i\xi)e^{ix\xi}d\xi, \qquad x \geq 0.\end{aligned}$$

With β fixed, define the complex variable $s = \beta + i\xi$, so that $ds = id\xi$ and the integral over the interval $(-L, L)$ becomes a contour integral in the complex s-plane over the line segment from $\beta - iL$ to $\beta + iL$. Thus

$$f(x)e^{-\beta x} = \lim_{L\to\infty} \frac{1}{2\pi i}\int_{\beta-iL}^{\beta+iL} F(s)e^{x(s-\beta)}ds,$$

and

$$f(x) = \lim_{L\to\infty} \frac{1}{2\pi i}\int_{\beta-iL}^{\beta+iL} F(s)e^{xs}ds, \tag{7.4}$$

which is the desired inversion formula for the Laplace transformation $\mathcal{L} : f \mapsto F$. Symbolically, we can write (7.4) as

$$f(x) = \mathcal{L}^{-1}(F)(x),$$

where

$$\mathcal{L}^{-1}(F)(x) = \lim_{L\to\infty} \frac{1}{2\pi i} \int_{\beta-iL}^{\beta+iL} F(s)e^{xs}ds$$

is the inverse Laplace transform of F.

One should keep in mind that (7.4) is a pointwise equality. It does not apply, for example, to functions which have singularities in $[0,\infty)$, for these are excluded by the assumption that f is piecewise smooth. Furthermore, because $f(0^-) = 0$, we always have

$$\lim_{L\to\infty} \frac{1}{2\pi i} \int_{\beta-iL}^{\beta+iL} F(s)ds = \frac{1}{2}f(0^+). \tag{7.5}$$

The contour integral in (7.4) is not always easy to evaluate by direct integration. On the contrary, this formula is used to evaluate the integral when the function f is known. The same observation applies to (7.5). For example, with reference to Example 7.3, if $\varepsilon > 0$,

$$\lim_{L\to\infty} \frac{1}{2\pi i} \int_{\varepsilon-iL}^{\varepsilon+iL} \frac{1}{s}e^{xs}ds = \begin{cases} 0, & x<0 \\ 1/2, & x=0 \\ 1, & x>0. \end{cases}$$

The inversion formula clearly implies that the Laplace transformation $\mathcal{L}$ is injective on piecewise smooth functions in $\mathcal{E}$. Thus, if $F = \mathcal{L}(f)$ and $G = \mathcal{L}(g)$, where f and g are piecewise smooth functions in $\mathcal{E}$, then

$$F = G \Rightarrow f = g.$$

EXERCISES

7.1 Determine the Laplace transform $F(s)$ of each of the following functions defined on $(0,\infty)$.

(a) $f(x) = (ax+b)^2$

(b) $f(x) = \cosh x$

(c) $f(x) = \sin^2 x$

(d) $f(x) = \sin x \cos x$

(e) $f(x) = \begin{cases} c, & 0<x<a \\ 0, & a<x<\infty \end{cases}$

(f) $f(x) = \begin{cases} a - ax/b, & 0<x<b \\ 0, & x>b \end{cases}$

(g) $f(x) = x \sinh x$

(h) $f(x) = x^2 e^x$

(i) $f(x) = 1/\sqrt{x}$.

7.2 Determine the inverse Laplace transform $f(x)$ for each of the following transform functions.

(a) $F(s) = \dfrac{a}{s+b}$

(b) $F(s) = \dfrac{2s-5}{s^2-9}$

(c) $F(s) = \dfrac{1}{s(s+1)}$

(d) $F(s) = \dfrac{1}{s^2+2s}$

(e) $F(s) = \dfrac{3(s-1)}{s^2-6}$

(f) $F(s) = 1/s^{3/2}$

(g) $F(s) = \dfrac{14s^2+55s+51}{2s^3+12s^2+22s+12}$.

7.3 If $a > 0$, prove that

$$\mathcal{L}(f(ax))(s) = \frac{1}{a}F(s/a),$$

where $F(s) = \mathcal{L}(f(x))(s)$.

7.2 Properties and Applications

As with the Fourier transform, when a function and its derivative have Laplace transforms, the transform functions are related by a formula which is easily derived by using integration by parts. We should recall at the outset that a piecewise continuous function is always locally integrable. Hence all piecewise continuous functions of exponential growth lie in $\mathcal{E}$.

Theorem 7.6

(i) If f is a continuous and piecewise smooth function on $[0, \infty)$ such that both $e^{-\alpha x} f(x)$ and $e^{-\alpha x} f'(x)$ are bounded, then

$$\mathcal{L}(f')(s) = s\mathcal{L}(f)(s) - f(0^+), \qquad \operatorname{Re} s > \alpha. \tag{7.6}$$

(ii) If f is a piecewise continuous function on $[0,\infty)$ and $e^{-\alpha x}f(x)$ is bounded, then

$$\mathcal{L}\left(\int_0^x f(t)dt\right)(s) = \frac{1}{s}\mathcal{L}(f)(s), \qquad x > 0, \qquad \operatorname{Re} s > \alpha. \tag{7.7}$$

Proof

(i) The assumptions on f guarantee the existence of both $\mathcal{L}(f)(s)$ and $\mathcal{L}(f')(s)$ on $\operatorname{Re} s > \alpha$. Integrating by parts,

$$\begin{aligned}\mathcal{L}(f')(s) &= \int_0^\infty f'(x)e^{-sx}dx \\ &= e^{-sx}f(x)\Big|_0^\infty + s\int_0^\infty f(x)e^{-sx}dx \\ &= s\mathcal{L}(f)(s) - f(0^+), \qquad \operatorname{Re} s > \alpha.\end{aligned}$$

Note that the continuity of f is needed for the second equality to be valid (see Exercise 6.9).

(ii) Let

$$g(x) = \int_0^x f(t)dt.$$

Because the function $|e^{-\alpha x}f(x)|$ is bounded on $[0,\infty)$ by some positive constant, say M, we have

$$|g(x)| \le M\int_0^x e^{\alpha t}dt \le \frac{M}{\alpha}(e^{\alpha x} - 1), \qquad x \ge 0, \qquad \alpha \ne 0,$$

hence

$$e^{-\alpha x}|g(x)| \le \frac{M}{|\alpha|}.$$

This implies that $e^{-\alpha x}g(x)$ is bounded on $[0,\infty)$. Now g is continuous and its derivative $g' = f$ is piecewise continuous on $[0,\infty)$, so we can use part (i) to conclude that

$$\mathcal{L}(f)(s) = \mathcal{L}(g')(s) = s\mathcal{L}(g) - g(0^+).$$

With $g(0^+) = 0$ we arrive at the desired result. The case where $\alpha = 0$ is straightforward. □

By induction on n, we can generalize (7.6) to arrive at the corresponding formula for the nth derivatives of f, and we leave the details of the proof to the reader.

Corollary 7.7

Let $f, f', \ldots, f^{(n-1)}$ be continuous and piecewise smooth functions defined on $[0, \infty)$. If $f, f', \ldots, f^{(n)}$ are of exponential growth, then $f^{(n)}$ has a Laplace transform on $\operatorname{Re} s > \alpha$, for some real number α, given by

$$\mathcal{L}(f^{(n)})(s) = s^n \mathcal{L}(f)(s) - s^{n-1} f(0^+) - s^{n-2} f'(0^+) - \cdots - f^{(n-1)}(0^+). \quad (7.8)$$

Example 7.8

(i) Using the result of Example 7.5 and Theorem 7.6,

$$\begin{aligned}\mathcal{L}(\cos ax)(s) &= \mathcal{L}\left(\frac{1}{a}\frac{d}{dx}\sin ax\right)(s) \\ &= \frac{1}{a}\left(s\frac{a}{s^2+a^2} - \sin 0\right) \\ &= \frac{s}{s^2+a^2}, \qquad s > 0.\end{aligned}$$

(ii) Given

$$\mathcal{L}(f)(s) = \frac{1}{s(s^2-1)}, \qquad s > 1,$$

we can apply the formula (7.7) to obtain

$$f(x) = \mathcal{L}^{-1}\left(\frac{1}{s}\frac{1}{s^2-1}\right)(x) = \int_0^x \sinh t dt = \cosh x - 1.$$

According to Theorem 7.6, differentiation of f is transformed under $\mathcal{L}$ to multiplication by s, followed by subtraction of $f(0^+)$, and integration over $(0, x)$ is transformed to division by s. Conversely, if $f \in \mathcal{E}$ and $\mathcal{L}(f) = F$, then we can show that

$$\mathcal{L}(xf)(s) = -\frac{d}{ds}\mathcal{L}(f)(s) = -F'(s) \quad (7.9)$$

$$\mathcal{L}(f/x)(s) = \int_s^\infty F(z)dz, \quad (7.10)$$

where it is assumed in (7.10) that the function $f(x)/x$ is locally integrable, and the contour of integration in the z-plane from s to ∞ is such that $\operatorname{Im} z$ remains bounded. We leave it to the reader to prove these two formulas.

7.2.1 Applications to Ordinary Differential Equations

Theorem 7.6 and its corollary make the Laplace transform an effective tool for solving initial-value problems for linear ordinary differential equations, especially when the coefficients in the equation are constant. It reduces the problem to first solving an algebraic equation and then taking the Laplace inverse transform of the solution. Consider, for example, the second-order differential equation

$$y'' + ay' + by = f(x), \qquad x > 0, \tag{7.11}$$

where a and b are constants, under the initial conditions

$$y(0^+) = y_0, \qquad y'(0^+) = y_1. \tag{7.12}$$

Using Corollary 7.7, with $\mathcal{L}(y) = Y$ and $\mathcal{L}(f) = F$, we obtain

$$s^2 Y - sy_0 - y_1 + a(sY - y_0) + bY = F.$$

Hence

$$Y(s) = \frac{F(s)}{s^2 + as + b} + \frac{y_0(s+a) + y_1}{s^2 + as + b},$$

and the solution is given by

$$\begin{aligned} y(x) &= \mathcal{L}^{-1}(Y)(x) \\ &= \mathcal{L}^{-1}\left[\frac{F(s)}{s^2 + as + b}\right](x) + \mathcal{L}^{-1}\left[\frac{y_0(s+a) + y_1}{s^2 + as + b}\right](x) \\ &= y_p(x) + y_h(x). \end{aligned}$$

y_p is, of course, a particular solution of the nonhomogeneous equation (7.11), and y_h is the corresponding homogeneous solution. This method provides a convenient way for obtaining y_p, especially when the function f is not continuous. The homogeneous part of the solution is simply the inverse transform of a rational function in s, which may be evaluated by using partial fractions, and which vanishes when the initial conditions on y and y' are homogeneous, that is, when $y_0 = y_1 = 0$. Thus the nonhomogeneous terms in either the differential equation or the boundary condition merely add more terms to the transform function Y, but otherwise pose no extra complication.

Equations (7.11) and (7.12) in fact represent a mathematical model for current flow in an RLC electric circuit. As shown in Figure 7.2, this consists of a resistance, measured by R, an inductance L, and a capacitance C, connected in series and supplied with an input voltage $v(t)$, which depends on time t. If the current in the circuit is denoted $y(t)$, then the drop in potential across each of the three circuit elements is given, respectively, by

$$v_R(t) = Ry(t), \qquad v_L(t) = Ly'(t), \qquad v_C = \frac{1}{C}\int_{t_0}^{t} y(\tau)d\tau.$$

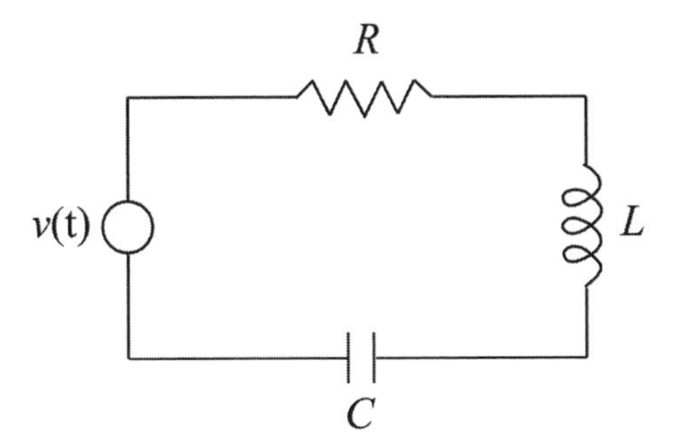

Figure 7.2 RLC series circuit.

According to Kirchoff's law, the sum of these potential drops is equal to the input voltage,

$$Ry + Ly' + \frac{1}{C}\int_{t_0}^{t} y(\tau)d\tau = v(t).$$

By differentiating with respect to t and dividing by L, we end up with the second-order equation

$$y'' + \frac{R}{L}y' + \frac{1}{LC}y = f(t),$$

where the coefficients are constant and $f = v'/L$ is given. If y and y' are specified at $t = 0$, when the circuit switch is closed, then we have a system of equations similar to Equations (7.11) and (7.12). The current is determined by solving the system along the lines explained above.

Example 7.9

Solve the initial-value problem

$$y'' + 4y' + 6y = e^{-t}, \qquad t > 0,$$
$$y(0) = 0, \qquad y'(0) = 0.$$

Solution

The Laplace transform of the differential equation is

$$s^2Y + 4sY + 6Y = \frac{1}{s+1},$$

and the transform function is

$$Y(s) = \frac{1}{(s+1)(s^2+4s+6)} = \frac{1}{(s+1)[(s+2)^2+2]}.$$

To invert this rational function we use partial fractions to write

$$Y(s) = \frac{1}{3}\frac{1}{s+1} - \frac{1}{3}\frac{s+2}{(s+2)^2+2} - \frac{1}{3}\frac{1}{(s+2)^2+2}.$$

By the linearity of the inverse transformation, we therefore have

$$y(t) = \frac{1}{3}e^{-t} - \frac{1}{3}\mathcal{L}^{-1}\left[\frac{s+2}{(s+2)^2+2}\right](t) - \frac{1}{3}\mathcal{L}^{-1}\left[\frac{1}{(s+2)^2+2}\right](t). \tag{7.13}$$

To evaluate the last two expressions, we need to determine the effect of translation in s on the inverse Laplace transform $\mathcal{L}^{-1}$. This is provided by the following translation theorem.

Theorem 7.10

If $f \in \mathcal{E}$ and $\mathcal{L}(f)(s) = F(s)$ on $\operatorname{Re} s > \alpha$, then

$$\mathcal{L}(e^{ax}f)(s) = F(s-a), \qquad s - a > \alpha, \tag{7.14}$$

$$\mathcal{L}[H(x-a)f(x-a)](s) = e^{-as}F(s), \qquad a \geq 0,\ s > \alpha. \tag{7.15}$$

Proof

Equation (7.14) is a direct result of the definition of the Laplace transform,

$$\begin{aligned}\mathcal{L}(e^{ax}f)(s) &= \int_0^\infty f(x)e^{-(s-a)x}dx \\ &= F(s-a), \qquad s-a > \alpha.\end{aligned}$$

To prove (7.15), note that

$$\begin{aligned}e^{-as}F(s) &= \int_0^\infty f(x)e^{-s(x+a)}dx \\ &= \int_a^\infty f(x-a)e^{-sx}dx \\ &= \int_0^\infty H(x-a)f(x-a)e^{-sx}dx \\ &= \mathcal{L}[H(x-a)f(x-a)](s).\end{aligned}$$

□

Going back to Equation (7.13) in Example 7.9, since

$$\mathcal{L}^{-1}\left(\frac{s}{s^2+2}\right)(t) = \cos\sqrt{2}t$$

$$\mathcal{L}^{-1}\left(\frac{1}{s^2+2}\right)(t) = \frac{1}{\sqrt{2}}\mathcal{L}^{-1}\left(\frac{\sqrt{2}}{s^2+2}\right) = \frac{1}{\sqrt{2}}\sin\sqrt{2}t,$$

we can apply the formula (7.14) to obtain

$$\mathcal{L}^{-1}\left[\frac{s+2}{(s+2)^2+2}\right](t) = e^{-2t}\cos\sqrt{2}t$$

$$\mathcal{L}^{-1}\left[\frac{1}{(s+2)^2+2}\right](t) = \frac{1}{\sqrt{2}}e^{-2t}\sin\sqrt{2}t.$$

The solution (7.13) is therefore given by

$$y(t) = \frac{1}{3}e^{-t} - \left(\frac{1}{3}\cos\sqrt{2}t + \frac{1}{3\sqrt{2}}\sin\sqrt{2}t\right)e^{-2t}.$$

Remark 7.11

One may wonder why the second-order equation in the above example has to have two initial conditions at $x = 0$ to determine its solution, whereas a similar equation, $y - y'' = f$ with $0 \leq x < \infty$, which was discussed at the end of Chapter 6, required only one condition, namely $y(0) = y_0$. The reason is that this last equation was solved using the Fourier transformation under the tacit assumption that y is an $\mathcal{L}^1$ function which, being continuous, tends to 0 as $x \to \infty$. This constitutes a second boundary condition.

In Example 7.10 we could just as well have used the conventional method for solving differential equations with constant coefficients. But in the following example, where the nonhomogeneous term is not continuous, there is a clear advantage to using the Laplace transform.

Example 7.12

Solve the initial-value problem

$$y' + 3y = f(t), \qquad t > 0,$$

$$y(0) = 1,$$

where

$$f(t) = \begin{cases} 0, & t < 0 \\ t, & 0 < t < 1 \\ 0, & t > 1. \end{cases}$$

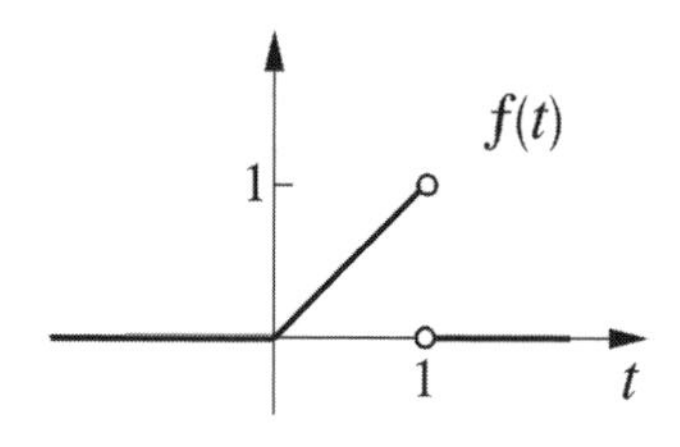

Figure 7.3

Here $y(t)$ represents the current in an RL circuit where the input voltage $f(t) = t$ is switched on at $t = 0$, and then switched off at $t = 1$.

Solution

By expressing f in the form $f(t) = t[H(t) - H(t-1)]$,
we can use the formula (7.9) and Theorem 7.10 to write

$$\begin{aligned}
\mathcal{L}(f)(s) &= -\frac{d}{ds}\mathcal{L}[(H(t) - H(t-1))](s) \\
&= -\frac{d}{ds}\left[\frac{1}{s} - \frac{e^{-s}}{s}\right] \\
&= \frac{1}{s^2} - \frac{e^{-s}}{s^2} - \frac{e^{-s}}{s}.
\end{aligned}$$

The transformed equation therefore has the form

$$\begin{aligned}
(s+3)Y(s) - 1 &= \frac{1}{s^2} - \frac{e^{-s}}{s^2} - \frac{e^{-s}}{s} \\
Y(s) &= \frac{1}{s+3} + \frac{1}{s^2}\frac{1}{s+3} - e^{-s}\frac{1}{s^2}\frac{1}{s+3} - e^{-s}\frac{1}{s}\frac{1}{s+3}.
\end{aligned}$$

With

$$\mathcal{L}^{-1}\left(\frac{1}{s+3}\right) = e^{-3t}H(t),$$

we use (7.7) to write

$$\begin{aligned}
\mathcal{L}^{-1}\left(\frac{1}{s}\frac{1}{s+3}\right)(t) &= \int_0^t e^{-3\tau}d\tau = \frac{1}{3}(1 - e^{-3t})H(t) \\
\mathcal{L}^{-1}\left(\frac{1}{s^2}\frac{1}{s+3}\right)(t) &= \int_0^t \frac{1}{3}(1 - e^{-3\tau})d\tau = \frac{1}{3}\left(t + \frac{1}{3}e^{-3t} - \frac{1}{3}\right)H(t).
\end{aligned}$$

Now (7.15) implies

$$\begin{aligned}
\mathcal{L}^{-1}\left(e^{-s}\frac{1}{s}\frac{1}{s+3}\right)(t) &= \frac{1}{3}[1 - e^{-3(t-1)}]H(t-1) \\
\mathcal{L}^{-1}\left(e^{-s}\frac{1}{s^2}\frac{1}{s+3}\right)(t) &= \frac{1}{3}\left[(t-1) + \frac{1}{3}e^{-3(t-1)} - \frac{1}{3}\right]H(t-1),
\end{aligned}$$

hence

$$\begin{aligned} y(t) &= \frac{1}{3}\left(t + \frac{10}{3}e^{-3t} - \frac{1}{3}\right)H(t) - \frac{1}{3}\left(t - \frac{2}{3}e^{-3(t-1)} - \frac{1}{3}\right)H(t-1) \\ &= \begin{cases} \frac{1}{3}\left(t - \frac{1}{3}\right) + \frac{10}{9}e^{-3t}, & 0 < t \leq 1 \\ \frac{10 + 2e^3}{9}e^{-3t}, & t > 1. \end{cases} \end{aligned}$$

Although f is discontinuous at $x = 1$, the solution y is continuous on $(0, \infty)$. This is to be expected, because the inductance L does not allow sudden changes in the current in an RL circuit.

The method used to arrive at the solutions in Examples 7.8, 7.9, and 7.12 forms part of a collection of techniques based on the properties of the Laplace transformation, known as the operational calculus, which was developed by the English physicist Oliver Heaviside (1850–1925) to solve linear differential equations of electric circuits. This method is not limited to equations with constant coefficients, as the next example shows.

Example 7.13

Under the Laplace transformation, Laguerre's equation

$$xy'' + (1 - x)y' + ny = 0, \qquad x > 0, \qquad n \in \mathbb{N}_0,$$

becomes

$$\begin{gathered} -\frac{d}{ds}[s^2Y - sy(0) - y'(0)] + sY - y(0) + \frac{d}{ds}[sY - y(0)] + nY = 0 \\ (s - s^2)Y' + (n + 1 - s)Y = 0 \\ \frac{Y'}{Y} = \frac{n + 1 - s}{s(s-1)} = \frac{n}{s-1} - \frac{n+1}{s}. \end{gathered}$$

Integrating the last equation, we obtain

$$Y(s) = c\frac{(s-1)^n}{s^{n+1}},$$

where c is the integration constant. Taking the inverse transform,

$$\begin{aligned} y(x) &= c\mathcal{L}^{-1}\left[\frac{(s-1)^n}{s^{n+1}}\right](x) \\ &= ce^x\mathcal{L}^{-1}\left[\frac{s^n}{(s+1)^{n+1}}\right](x) \\ &= \frac{c}{n!}e^x\frac{d^n}{dx^n}(e^{-x}x^n). \end{aligned}$$

Laguerre's polynomial L_n is obtained by setting $c = 1$.

7.2.2 The Telegraph Equation

The Laplace transform can also be used to solve boundary-value problems for partial differential equations, especially where the time variable t is involved; for then the transform can usually be applied with respect to t over the semi-infinite interval $[0,\infty)$. Consider the equation

$$u_{xx} = Au_{tt} + Bu_t + Cu, \qquad 0 < t < \infty, \tag{7.16}$$

where A, B, and C are nonnegative constants. This equation, known as the telegraph equation, describes an electromagnetic signal $u(x,t)$, such as an electric current or voltage, traveling along a transmission line. The constants A, B, C are determined by the distributed inductance, resistance, and capacitance (per unit length) along the line (see [7], vol. 2). If the transmission line extends over $-\infty < x < \infty$, then two initial conditions (at $t = 0$) on u and u_t are sufficient to specify u. But if it extends over $0 \leq x < \infty$, then we need to specify u at $x = 0$ as well, in which case the boundary conditions take the form

$$\begin{aligned} u(0,t) &= f(t), & t &\geq 0, \\ u(x,0) &= g(x), & x &\geq 0, \\ u_t(x,0) &= h(x), & x &\geq 0. \end{aligned}$$

Here $f(t)$ is the input signal which is transmitted, and the two initial conditions are needed because Equation (7.16) is of second order in t (if $A = 0$ we do not need the condition on u_t). We can attempt to solve Equation (7.16) by separation of variables, using the Fourier integral as we did previously, but the fact that all the boundary conditions are nonhomogeneous makes the procedure more difficult. It also places unnecessary restrictions on the behaviour of u and f as $t \to \infty$.

We can simplify things by solving Equation (7.16) under two sets of boundary conditions: first with $f = 0$, and then with $g = h = 0$. Because (7.16) is linear and homogeneous, the sum of these two solutions is also a solution of the equation, and it satisfies the three nonhomogeneous boundary conditions. In both cases, the Laplace transform can be used to construct the solution. Let us take the second case, where the boundary conditions are

$$\begin{aligned} u(0,t) &= f(t), \qquad t \geq 0, \\ u(x,0) &= u_t(x,0) = 0, \qquad x \geq 0. \end{aligned}$$

Assuming that f, u, u_x, u_{xx}, u_t, and u_{tt} all lie in $\mathcal{E}$ as functions of t, and that u and its first partial derivatives are piecewise smooth functions of t, we can apply the Laplace transformation with respect to t to Equation (7.16) to obtain

$$U_{xx} = (As^2 + Bs + C)U, \tag{7.17}$$

where

$$U(x, s) = \int_0^\infty u(x, t)e^{-st} dt \tag{7.18}$$

and $\operatorname{Re} s$ is greater than some real number α. If, for example, the signal u is bounded we may take $\alpha = 0$. Note that differentiation with respect to x may be taken inside the integral (7.18) because the latter converges uniformly on $\operatorname{Re} s \geq \alpha + \varepsilon$ for any $\varepsilon > 0$. For every fixed value of the parameter s, Equation (7.17) has the general solution

$$U(x, s) = c_1(s)e^{\lambda(s)x} + c_2(s)e^{-\lambda(s)x},$$

where $c_1 + c_2 = U(0, s) = \mathcal{L}(f)(s) = F(s)$ and $\lambda(s) = \sqrt{As^2 + Bs + C}$. If we take the principal branch of the square root, where $\operatorname{Re} \lambda(s) > 0$, the solution $e^{\lambda(s)x}$ becomes unbounded in the right half of the s-plane and must therefore be dropped (by setting $c_1 = 0$). The resulting transform function

$$U(x, s) = F(s)e^{-\lambda(s)x}$$

can, in principle, be inverted to yield the desired solution $u(x, t) = \mathcal{L}^{-1}(U)(x, t)$, although the computations involved may be quite tedious. Nevertheless, there are a number of significant special cases of Equation (7.16) where the solution can be obtained explicitly.

(i) If $A = 1/c^2$ and $B = C = 0$, (7.16) reduces to the wave equation

$$u_{tt} = c^2 u_{xx}.$$

Here the transform function

$$U(x, s) = e^{-sx/c} F(s)$$

is easily inverted, using formula (7.15), to

$$u(x, t) = H(t - x/c)f(t - x/c).$$

This expression indicates that there is no signal at a point x until $t = x/c$. Thus the signal $f(t)$ moves down the line without distortion or damping and with velocity c. This is to be expected, because the damping factor Bu_t is absent and, consequently, no energy is dissipated in the transmission.

(ii) If $As^2 + Bs + C = (s + b)^2/c^2$ is a perfect square, then

$$U(x, s) = e^{-(s+b)x/c} F(s)$$

and the solution is

$$u(x,t) = e^{-bx/c}H(t - x/c)f(t - x/c).$$

Here the signal $f(t)$ again moves down the line with velocity c, undistorted, but attenuated by the damping factor $e^{-bx/c}$. A transmission line therefore becomes distortionless if it is designed so that the distributed resistance, inductance, and capacitance are such that the relation $B^2 = 4AC$ holds. The interested reader may refer to [10] on this point and on some other interesting historical background to the telegraph equation.
(iii) If $A = C = 0$ and $B = 1/k$, where k is a positive constant, we obtain the heat equation

$$u_t = ku_{xx}.$$

In this case the Laplace transform of u is

$$U(x,s) = e^{-x\sqrt{s/k}}F(s), \tag{7.19}$$

which is not readily invertible. To determine $\mathcal{L}^{-1}(U)$ we need another rule which allows us to invert the product of two transforms. By analogy with the Fourier transformation we should expect this to involve the convolution of the inverse transforms.

If the functions f and g are locally integrable on $[0,\infty)$, then their convolution

$$f * g(x) = \int_0^x f(x-t)g(t)dt = \int_0^x f(t)g(x-t)dt = g * f(x)$$

is well defined and locally integrable on $[0,\infty)$. For if $[a,b]$ is any finite interval in $[0,\infty)$, then

$$\begin{aligned}\int_a^b |f * g(x)|\,dx &\le \int_a^b \int_0^x |f(x-t)|\,|g(t)|\,dtdx \\ &= \int_a^b \int_0^b H(x-t)\,|f(x-t)|\,|g(t)|\,dtdx \\ &= \int_0^b \left[\int_a^b H(x-t)\,|f(x-t)|\,dx\right]|g(t)|\,dt \\ &\le \int_0^b \left[\int_{a-t}^{b-t} H(y)\,|f(y)|\,dy\right]|g(t)|\,dt.\end{aligned}$$

Because f is locally integrable and $0 \le t \le b$, the integral of $H|f|$ over $[a-t, b-t]$ is uniformly bounded by a constant, and, because g is also locally integrable, $|f * g|$ is integrable on $[a,b]$. The convolution $f * g$ is also (piecewise) continuous if either f or g is (piecewise) continuous, and (piecewise) smooth if either f or g is (piecewise) smooth (Exercise 7.16).

If $f(x)$ and $g(x)$ are dominated as $x \to \infty$ by $e^{\alpha x}$, then one can easily check that $f * g(x)$ will be dominated by $e^{\beta x}$ for any $\beta > \alpha$. Consequently, if f and g belong to $\mathcal{E}$, then so does their convolution $f * g$, and its Laplace transform is given by

$$\begin{aligned}\mathcal{L}(f * g)(s) &= \int_0^\infty e^{-sx} \int_0^x f(t)g(x-t)dtdx \\ &= \int_0^\infty \int_0^\infty H(x-t)f(t)g(x-t)e^{-sx}dtdx \\ &= \int_0^\infty f(t) \int_0^\infty g(y)e^{-s(t+y)}dydt \\ &= \mathcal{L}(f)\mathcal{L}(g)(s).\end{aligned}$$

In the third equality, the order of integration is reversed, and this is justified by the uniform convergence of the double integral on $\operatorname{Re} s \geq \beta + \varepsilon$ for any positive ε. Thus we have proved the following convolution theorem which corresponds to Theorem 6.22 for the Fourier transformation.

Theorem 7.14

Let $f, g \in \mathcal{E}$. If $\mathcal{L}(f)(s) = F(s)$ and $\mathcal{L}(g)(s) = G(s)$, then

$$\mathcal{L}(f * g)(s) = F(s)G(s).$$

Now we can go back to Equation (7.19) to conclude that

$$\begin{aligned}u(x,t) &= \mathcal{L}^{-1}\left(e^{-x\sqrt{s/k}}F(s)\right)(t) \\ &= f * \mathcal{L}^{-1}\left(e^{-x\sqrt{s/k}}\right)(t).\end{aligned}$$

The function $\mathcal{L}^{-1}(e^{-x\sqrt{s/k}})(t)$ can be evaluated by using the inversion formula (7.4), which requires some manipulations of contour integrals (see Exercise 7.21), or it may be looked up in a table of Laplace transforms. In either case

$$\mathcal{L}^{-1}\left(e^{-x\sqrt{s/k}}\right)(t) = \frac{x}{\sqrt{4\pi k t^3}} e^{-x^2/4kt},$$

hence

$$u(x,t) = \frac{x}{\sqrt{4\pi k}} \int_0^t f(t-\tau)\tau^{-3/2}e^{-x^2/4k\tau}d\tau. \tag{7.20}$$

Here the solution u differs considerably from that in the first two cases. It tends to 0 as $x \to \infty$ at any time t and also as $t \to \infty$ at any point x, and the signal

f is distorted as it spreads along the line with no fixed velocity. This is not surprising, as the original equation (7.16) is radically changed by dropping the first term. It changes from a hyperbolic equation, with wavelike solutions, to a parabolic equation whose solutions travel in a diffusive manner, in much the same way that heat or gas spreads out.

EXERCISES

7.4 Sketch each of the following functions and determine its Laplace transform.

(a) $(x-1)H(x-1)$

(b) $(x-1)^2 H(x-1)$

(c) $x^2[H(x-1)-H(x-3)]$

(d) $H(x-\pi/2)\cos x$

(e) $(1-e^{-x})[H(x)-H(x-1)]$.

7.5 Use the Heaviside function to represent the function

$$f(x)=\begin{cases} x, & 0<x<1 \\ e^{1-x}, & x>1 \end{cases}$$

by a single expression. Sketch the function and determine its Laplace transform.

7.6 Determine the inverse Laplace transform of each of the following functions.

(a) $\dfrac{e^{-6s}}{s^3}$

(b) $\dfrac{e^{-s}}{s^2+2s+2}$

(c) $\dfrac{1}{s}(e^{-3s}+e^{-s})$

(d) $\dfrac{1}{s-1}(e^{-3s}+e^{-s})$

(e) $\dfrac{1+e^{-\pi s}}{s^2+9}$.

7.7 If f is not assumed to be continuous in Theorem 7.6(i), what would Equation (7.6) look like?

7.8 Solve the following initial-value problems:

(a) $y'' + 4y' + 5y = 0,\ y(0) = 1,\ y'(0) = 1$

(b) $9y'' - 6y' + y = 0,\ y(0) = 3,\ y'(0) = 1$

(c) $y'' + 2y' + 5y = 3e^{-x}\sin x,\ y(0) = 0,\ y'(0) = 3$

(d) $y'' + 2y' - 8y = -256x^3,\ y(0) = 15,\ y'(0) = 36$

(e) $y'' - 3y' + 2y = H(t-1),\ y(0) = 0,\ y'(0) = 1$

(f) $y' + 2y = x[H(x) - H(x-1)]$

(g) $y' + y = \begin{cases} \sin x, & 0 < x < \pi \\ -2\sin x, & x > \pi \end{cases},\ y(0) = 0,\ y'(0) = 0.$

7.9 Invert each of the following transforms.

(a) $\dfrac{s}{(s^2+9)^2}$

(b) $\log\left(\dfrac{s}{s-1}\right)$

(c) $\log\left(\dfrac{s+a}{s+b}\right)$

(d) $\cot^{-1}(s+1)$.

7.10 The function $\mathrm{Si} : \mathbb{R} \to \mathbb{R}$, defined by the improper integral

$$\mathrm{Si}(x) = \int_0^x \frac{\sin t}{t}\,dt,$$

is called the sine integral. Prove that

$$\mathcal{L}\left(\frac{\sin x}{x}\right)(s) = \tan^{-1}\left(\frac{1}{s}\right)$$

$$\mathcal{L}[\,\mathrm{Si}(x)](s) = \frac{1}{s}\tan^{-1}\left(\frac{1}{s}\right).$$

7.11 (a) Let $f \in \mathcal{E}$ be a periodic function on $[0, \infty)$ of period $p > 0$. Show that

$$\mathcal{L}(f)(s) = \frac{1}{1 - e^{-ps}} \int_0^p f(x)e^{-sx}\,dx, \qquad s > 0.$$

(b) Use this to compute $\mathcal{L}(f)$, where $f(x) = x$ on $(0, 1)$, $f(x+1) = f(x)$ for all $x > 0$, and $f(x) = 0$ for all $x < 0$.

7.12 Prove that $\mathcal{L}(f)(s) \to 0$ as $\mathrm{Re}\, s \to \infty$ for any $f \in \mathcal{E}$.

7.13 Use the Laplace transformation to solve the integral equation $\int_0^x (x-t)^3 y(t)dt = f(x)$, and state the conditions on f which ensure that the method works.

7.14 If $f(x) = e^x$ and $g(x) = 1/\sqrt{\pi x}$ prove that

$$f * g(x) = e^x \operatorname{erf}(\sqrt{x}).$$

Use this to evaluate $\mathcal{L}[e^x \operatorname{erf}(\sqrt{x})]$ and $\mathcal{L}[\operatorname{erf}(\sqrt{x})]$.

7.15 Determine $\mathcal{L}([x])$, where $[x]$ is the integer part of the nonnegative real number x; that is,

$$[x] = n \qquad \text{for all } x \in [n, n+1), \qquad n \in \mathbb{N}_0.$$

7.16 Given two locally integrable functions $f, g : [0, \infty) \to \mathbb{C}$, show that

(a) $f * g$ is continuous (smooth) if either f or g is continuous (smooth).

(b) $f * g$ is piecewise continuous (smooth) if either f or g is piecewise continuous (smooth).

7.17 Use the Laplace transformation to obtain the solution of the wave equation

$$u_{tt} = c^2 u_{xx}, \qquad x > 0,\ t > 0,$$

subject to the boundary condition

$$u(0,t) = \cos^2 t, \qquad t \geq 0,$$

and the initial conditions

$$u(x,0) = u_t(x,0) = 0, \qquad 0 \leq x \leq l.$$

7.18 Solve the boundary-value problem

$$\begin{aligned} u_t - au &= ku_{xx}, \qquad x > 0,\ t > 0, \\ u(0,t) &= f(t), \qquad t \geq 0, \\ u(x,0) &= 0, \qquad x \geq 0. \end{aligned}$$

7.19 Solve the boundary-value problem

$$\begin{aligned} u_{tt} + 2u_t + u &= c^2 u_{xx}, \qquad x > 0,\ t > 0, \\ u(0,t) &= \sin t, \qquad t \geq 0, \\ u(x,0) &= u_t(x,0) = 0, \qquad x \geq 0. \end{aligned}$$

7.20 In Equation (7.20), prove that $u(x,t) \to f(t)$ as $x \to 0$.

7.21 Use contour integration in the s-plane to prove that

$$\mathcal{L}^{-1}(e^{-a\sqrt{s}}/\sqrt{s})(x) = \frac{1}{\sqrt{\pi x}} e^{-a^2/4x},$$

where $a > 0$. Conclude from this formula that

$$\mathcal{L}^{-1}(e^{-a\sqrt{s}}) = \frac{a}{2\sqrt{\pi x^3}} e^{-a^2/4x}.$$

Solutions to Selected Exercises

Chapter 1

1.2 (a) Complex vector space, (b) real vector space, (c) not a vector space, (d) real vector space.

1.4 Assume that the numbers are different and show that this leads to a contradiction. Assuming $\{x_1, \ldots, x_n\}$ and $\{y_1, \ldots, y_{n+1}\}$ are bases of the same vector space, express each y_i, $0 \leq i \leq n$, as a linear combination of $x_1, \ldots, x_n$. The resulting system of n linear equations can be solved uniquely for each x_i, $0 \leq i \leq n$, as a linear combination of y_i, $0 \leq i \leq n$ (why?). Since y_{n+1} is also a linear combination of $x_1, \ldots, x_n$ (and hence of $y_1, \ldots, y_n$), this contradicts the linear independence of $y_1, \ldots, y_{n+1}$.

1.7 Recall that a determinant is zero if, and only if, one of its rows (or columns) is a linear combination of the other rows (or columns).

1.8 Use the equality $\|\mathbf{x}+\mathbf{y}\|^2 = \|\mathbf{x}\|^2 + 2\,\mathrm{Re}\,\langle \mathbf{x}, \mathbf{y}\rangle + \|\mathbf{y}\|^2$. Consider $\mathbf{x} = (1, 1)$ and $\mathbf{y} = (i, i)$.

1.10 (a) 0, (b) 2/3, (c) 8/3, (d) $\sqrt{14}$.

1.12 $\langle f, f_1\rangle / \|f_1\| = \sqrt{\pi/2}$, $\langle f, f_2\rangle / \|f_2\| = 0$, $\langle f, f_3\rangle / \|f_3\| = \sqrt{\pi}/2$.

1.14 No, because $|x| = x$ on $[0, 1]$..

1.17 $a = -1$, $b = 1/6$.

1.21 Use the definition of the Riemann integral to show that f is not integrable.

1.22 (i) $1/\sqrt{2}$, (ii) not in $\mathcal{L}^2$, (iii) 1, (iv) not in $\mathcal{L}^2$

1.23 Examine the proof at the beginning of Section 1.3.

1.24 Use Exercise 1.23 to show that f and g must be linearly dependent. Assuming $g = \alpha f$ show that $\alpha \geq 0$.

1.25 $\alpha > -1/2$.

1.26 $\alpha < -1/2$.

1.28 Use the CBS inequality. $f(x) = 1/\sqrt{x}$ on $(0, 1]$.

1.30 $\sin^3 x = \frac{3}{4}\sin x - \frac{1}{4}\sin 3x$.

1.32 Use the fact that, for any polynomial p, $p(x)e^{-x} \to 0$ as $x \to \infty$.

1.34 (a) The limit is the discontinuous function

$$\lim_{n\to\infty}\frac{x^n}{1+x^n} = \begin{cases} 1, & |x| > 1 \\ 0, & |x| < 1 \\ 1/2, & x = 1 \\ \text{undefined}, & x = -1 \end{cases}$$

1.35 (a) Pointwise, (b) uniform, (c) pointwise.

1.37 Pointwise convergent to

$$f(x) = \begin{cases} 0, & x = 0 \\ 1 - x, & 0 < x \leq 1. \end{cases}$$

1.40 The domain of definition of f_n is not bounded.

1.42 (a) $\mathbb{R}$, (b) $x \neq -1$.

1.50 (a) 1, (b) 1, (c) 0.

1.51 (a) Convergent, (b) convergent, (c) divergent.

1.57 $c_1 = 1$, $c_2 = -2/\pi$, $c_3 = -1/\pi$.

1.58 $a_0 = \pi/2$, $a_1 = -4/\pi$, $a_2 = 0$, $b_1 = 0$, $b_2 = 0$.

1.61 $a_k = 1/n$. $\mathcal{L}^2$ convergence.

Chapter 2

2.1 (a) $y = e^{2x}(c_1 \cos\sqrt{3}x + c_2 \sin\sqrt{3}x) + e^x/4$.

(b) $y = c_1 x^2 + c_2 + x^3$.

(c) $y = x^{-1}(c_1 + c_2 \log x) + \frac{1}{4}x - 1$.

2.3 $c_{n+2} = -\frac{2}{n+1}c_n$, $y = c_0\left(1 - 2x^2 + \frac{4}{3}x^4 + \cdots\right) + c_1\left(x - x^3 + \frac{1}{2}x^5 + \cdots\right)$, $|x| < 1$.

2.5 A second-order equation has at most two linearly independent solutions.

2.7 (a) $y'' + 2y' + 5y = 0$, (b) $x^2y'' + xy' - y = 0$, (c) $xy'' + y' = 0$.

2.8 Use Lemma 2.7 and the fact that a bounded infinite set of real numbers has at least one cluster (or limit) point. This property of the real (as well as the complex) numbers is known as the Bolzano–Weierstass theorem (see [1]).

2.10 Use Theorem 2.10.

2.12 The solutions of (a) and (c) are oscillatory.

2.13 $y = x^{-1/2}(c_1\cos x + c_2\sin x)$. The zeros of $x^{-1/2}\cos x$ are $\{\frac{\pi}{2} + n\pi\}$, and those of $x^{-1/2}\sin x$ are $\{n\pi\}$, $n \in \mathbb{Z}$.

2.15 Use Theorem 2.10.

2.17 (a) $e^{\pm\sqrt{\lambda}x}$, $\lambda \in \mathbb{C}$, (b) $e^{-\sqrt{\lambda}x}$, $\operatorname{Re}\sqrt{\lambda} > 0$.

2.19 (a) $\rho = 1/x^2$, (c) $\rho = e^{-x^3/3}$.

2.21 $\rho = e^{2x}$. $\lambda_n = n^2\pi^2 + 1$, $u_n(x) = e^{-x}\sin\lambda_n x$.

2.23 $\lambda_n = n^2\pi^2(b-a)^{-2}$, $u_n(x) = \sin\left(\dfrac{n\pi(x-a)}{b-a}\right)$.

2.27 (a), (b), (c), and (f).

2.29 Change the independent variable to $\xi = x+3$ and solve. $\lambda_n = \left(\dfrac{n\pi}{\log 4}\right)^2 + \dfrac{1}{4}$, $y_n(x) = (x+3)^{-1/2}\sin\left(\dfrac{n\pi}{\log 4}\log(x+3)\right)$.

2.31 Refer to Example 2.17.

2.32 Multiply by $\bar{u}$ and integrate over $[a, b]$.

Chapter 3

3.2 No, because its sum is discontinuous at $x = 0$ (Example 3.4).

3.4 $\pi - |x| = \dfrac{\pi}{2} + \dfrac{4}{\pi}\sum_{n=0}^{\infty}\dfrac{1}{(2n+1)^2}\cos(2n+1)x$. Uniformly convergent by the Weierstrass M-test.

3.6 Use the M-test.

3.9 (a) and (c) are piecewise continuous; (b), (d), and (e) are piecewise smooth.

3.11 Use the definition of the derivative at $x = 0$ to show that $f'(0)$ exists, then show that $\lim_{x\to 0^+} f'(x)$ does not exist.

3.15 (a) $S(x) = 2\sum_{n=1}^{\infty} \frac{(-1)^{n+1}}{n} \sin nx.$

(d) $\cos^3 x = \frac{3}{4}\cos x + \frac{1}{4}\cos 3x.$

3.16 The convergence is uniform where f is continuous, hence in (b), (c), and (d).

3.17 In Exercise 3.15 (e), $S(\pm 2) = \frac{1}{2}(e^2 + e^{-2})$, and in (f) $S(\pm l) = 0$.

3.19 $\pi^2 = 8\sum_{n=0}^{\infty}(2n+1)^{-2}.$

3.21 $x^2 = \frac{\pi^2}{3} + 4\sum_{n=1}^{\infty}\frac{(-1)^n}{n^2}\cos nx$. Evaluate at $x = 0$ and $x = \pi$.

3.23 f' is an odd function which is periodic in π with $f'(x) = \cos x$ on $[0, \pi]$, hence $S(x) = \sum_{k=1}^{\infty} b_k \text{sin} kx$ where $b_k = \frac{2k}{\pi}\left[\frac{1+(-1)^k}{k^2-1}\right]$, $k > 1$, $b_1 = 0$. $S(n\pi) = 0$, $S(\frac{\pi}{2} + n\pi) = f'(\frac{\pi}{2} + n\pi) = 0$.

3.27 (a) $u(x,t) = \frac{3}{4}e^{-t}\text{sin}x - \frac{1}{4}e^{-9t}\text{sin}3x.$

(b) $u(x,t) = e^{-k\pi^2 t/4}\text{sin}\frac{\pi x}{2} - e^{-25k\pi^2 t/36}\text{sin}\frac{5\pi x}{6}.$

3.29 $u(x,t) = \sum_{n=1}^{\infty} a_n \text{sin}\frac{n\pi}{l}x \cos\frac{n\pi}{l}t$, $a_n = \frac{2}{l}\int_0^l x(l-x)\text{sin}\frac{n\pi}{l}x\, dx.$

3.31 Assume $u(x,t) = v(x,t) + \psi(x)$ where v satisfies the homogeneous wave equation with homogeneous boundary conditions at $x = 0$ and $x = l$. This leads to $\psi(x) = \frac{g}{2c^2}(x^2 - lx)$, and $v(x,t) = \sum_{n=1}^{\infty} a_n \text{sin}\frac{n\pi}{l}x\cos\frac{cn\pi}{l}t$, with $a_n = -\frac{2}{l}\int_0^l \psi(x)\text{sin}\frac{n\pi}{l}x\, dx.$

3.33 Assume $u(x,y,t) = v(x,y)w(t)$ and conclude that $w''/w = c^2\Delta v/v = -\lambda^2$ (separation constant). Hence $w(t) = A\cos\lambda t + B\sin\lambda t$. Assume $v(x,y) = X(x)Y(y)$, and use the given boundary conditions to conclude that

$$\lambda = \lambda_{mn} = \sqrt{\frac{n^2}{a^2} + \frac{m^2}{b^2}}\,\pi, \qquad m, n \in \mathbb{N},$$

$$X(x) = \sin\frac{n\pi}{a}x, \qquad Y(y) = \sin\frac{m\pi}{b}y,$$

$$u_{mn}(x,y,t) = (A_{mn}\text{cos}\lambda_{mn}ct + B_{mn}\text{sin}\lambda_{mn}ct)\text{sin}\frac{n\pi}{a}x\ \sin\frac{m\pi}{b}y.$$

Apply the initial conditions to the solution

$$u(x,y,t) = \sum_{n=1}^{\infty}\sum_{m=1}^{\infty} u_{mn}(x,y,t)$$

to evaluate the coefficients. This yields

$$A_{mn} = \frac{4}{ab}\int_0^a\int_0^b f(x,y)\sin\frac{n\pi}{a}x\sin\frac{m\pi}{b}y\,dxdy,$$

$$B_{mn} = \frac{4}{\lambda_{mn}ab}\int_0^a\int_0^b g(x,y)\sin\frac{n\pi}{a}x\sin\frac{m\pi}{b}y\,dxdy.$$

3.35 $u(x,y) = \left(\sinh\frac{3\pi}{2}\right)^{-1}\sin\frac{3\pi}{2}x\,\sinh\frac{3\pi}{2}y.$

3.37 (b) Use the fact that u must be bounded at $r = 0$ to eliminate the coefficients d_n.

(c) $u(r,\theta) = A_0 + \sum_{n=1}^{\infty}\left(\frac{R}{r}\right)^n (A_n \cos n\theta + B_n \sin n\theta).$

Chapter 4

4.3 From the recursion formula (4.7) with $k = 2j$, it follows that

$$\lim_{j\to\infty}\frac{\left|c_{2(j+1)}x^{2(j+1)}\right|}{\left|c_{2j}x^{2j}\right|} = x^2 < 1 \quad \text{for all } x \in (-1,1).$$

The same conclusion holds if $k = 2j+1$. Because $Q_0'(x) = (1-x^2)^{-1}$,

$$\lim_{x\to\pm1} p(x)Q_0'(x) = 1,$$

whereas

$$\lim_{x\to\pm1} p(x)Q_0(x) = 0.$$

4.5 The first two formulas follow from the fact that P_n is an even function when n is even, and odd when n is odd.

$$P_{2n}(0) = a_0 = \frac{(-1)^n(2n)!}{2^{2n}n!n!} = (-1)^n\frac{(2n-1)\cdots(3)(1)}{(2n)\cdots(4)(2)}.$$

4.7 Differentiate Rodrigues' formula for P_n and replace n by $n+1$ to obtain

$$P'_{n+1} = \frac{1}{2^n n!} \frac{d^n}{dx^n} \left[\left((2n+1)x^2 - 1 \right) \left(x^2 - 1 \right)^{n-1} \right],$$

then differentiate P_{n-1} and subtract. The first integral formula follows directly from (4.14) and the equality $P_n(\pm 1) = (\pm 1)^n$. The second results from setting $x = 1$.

4.11 (a) $1 - x^3 = P_0(x) - \frac{3}{5} P_1(x) - \frac{2}{5} P_3(x)$.

(b) $|x| = \frac{1}{2} P_0(x) + \frac{5}{8} P_2(x) - \frac{3}{16} P_4(x) + \cdots$.

4.13 $f(x) = \sum_{n=0}^{\infty} c_n P_n(x)$, where $c_n = (2n+1)/2 \int_{-1}^{1} f(x) P_n(x) dx$. Because f is odd, $c_n = 0$ for all even values of n. For $n = 2k+1$,

$$\begin{aligned} c_{2k+1} &= (4k+3) \int_0^1 P_{2k+1}(x) dx \\ &= (4k+3) \frac{1}{4k+3} [P_{2k}(0) - P_{2k+2}(0)] \\ &= (-1)^k \left[\frac{(2k)!}{2^{2k} k! k!} + \frac{(2k+2)!}{2^{2k+2}(k+1)!(k+1)!} \right] \\ &= (-1)^k \frac{(2k)!}{2^{2k} k! k!} \frac{(4k+3)}{(2k+2)}, \qquad k \in \mathbb{N}_0. \end{aligned}$$

Hence $f(x) = \frac{3}{2} P_1(x) - \frac{7}{8} P_3(x) + \frac{11}{16} P_5(x) + \cdots$. At $x = 0$,

$$\sum_{n=0}^{\infty} c_n P_n(0) = 0 = \frac{1}{2} [f(0^+) + f(0^-)].$$

4.15 $\left(\int_{-\infty}^{\infty} e^{-x^2} dx \right)^2 = 4 \int_0^{\infty} \int_0^{\infty} e^{-(x^2+y^2)} dx dy = 4 \int_0^{\infty} \int_0^{\pi/2} e^{r^2} r \, dr d\theta = \pi$.

4.16 Replace t by $-t$ in Equation (4.25) to obtain

$$\sum_{n=0}^{\infty} \frac{1}{n!} H_n(x)(-t)^n = e^{-2xt - t^2} = \sum_{n=0}^{\infty} \frac{1}{n!} H_n(-x) t^n,$$

which implies $H_n(-x) = (-1)^n H_n(x)$.

4.17 Setting $x = 0$ in (4.25) yields

$$\sum_{k=0}^{\infty} \frac{1}{k!} H_k(0) t^k = e^{-t^2} = \sum_{n=0}^{\infty} (-1)^n \frac{1}{n!} t^{2n}.$$

By equating corresponding coefficients we obtain the desired formulas.

4.19 If $m = 2n$,

$$x^{2n} = \frac{(2n)!}{2^{2n}} \sum_{k=0}^{n} \frac{H_{2k}(x)}{(2k)!(n-k)!}.$$

If $m = 2n+1$,

$$x^{2n+1} = \frac{(2n+1)!}{2^{2n+1}} \sum_{k=0}^{n} \frac{H_{2k+1}(x)}{(2k+1)!(n-k)!}, \qquad x \in \mathbb{R}, \qquad n \in \mathbb{N}_0.$$

4.23 Use Leibnitz' rule for the derivative of a product,

$$(fg)^{(n)} = \sum_{k=0}^{\infty} \binom{n}{k} f^{(n-k)} g^{(k)},$$

with $f(x) = x^n$ and $g(x) = e^{-x}$.

4.25 $x^m = \sum_{n=0}^{m} c_n L_n(x)$, where $c_n = \int_0^\infty e^{-x} x^m L_n(x) dx = (-1)^n \frac{m!m!}{n!(m-n)!}$.

4.28 $u(x) = c_1 + c_2 \int \frac{e^x}{x} dx = c_1 + c_2 \left(\log x + x + \frac{1}{2}\frac{x^2}{2!} + \cdots \right)$.

4.29 The surface $\varphi = \pi/2$, corresponding to the xy-plane.

4.31 The solution of Laplace's equation in the spherical coordinates (r, φ) is given by Equation (4.42). Using the given boundary condition in Equation (4.43),

$$\begin{aligned} a_n &= \frac{2n+1}{2R^n} \int_0^{\pi/2} 10 P_n(\cos\varphi) \sin\varphi \, d\varphi \\ &= \frac{5(2n+1)}{R^n} \int_0^1 P_n(x) dx \\ &= \frac{5}{R^n} [P_{n-1}(0) - P_{n+1}(0)], \qquad n \in \mathbb{N}, \end{aligned}$$

where the result of Exercise 4.7 is used in the last equality. We therefore arrive at the solution

$$\begin{aligned} u(r, \varphi) &= 5 + 5 \sum_{n=1}^{\infty} \left[P_{n-1}(0) - P_{n+1}(0) \right] \left(\frac{r}{R} \right)^n P_n(\cos\varphi) \Big] \\ &= 5 \left[1 + \frac{3}{2}\frac{r}{R} P_1(\cos\varphi) - \frac{7}{8} \left(\frac{r}{R} \right)^3 P_3(\cos\varphi) + \cdots \right]. \end{aligned}$$

Note that $u(R, \varphi) - 5$ is an odd function of φ, hence the summation (starting with $n = 1$) is over odd values of n.

4.33 In view of the boundary condition $u_\varphi(r, \pi/2) = 0$, f may be extended as an even function of φ from $[0, \pi/2]$ to $[0, \pi]$. By symmetry the solution is even about $\varphi = \pi/2$, hence the summation is over even orders of the Legendre polynomials.

Chapter 5

5.1 For all $n \in \mathbb{N}$, the integral $I_n(x) = \int_0^n e^{-t}t^{x-1}dt$ is a continuous function of $x \in [a,b]$, where $0 < a < b < \infty$. Because

$$0 \leq \int_n^\infty e^{-t}t^{x-1}dt \leq \int_n^\infty e^{-t}t^{b-1}dt \xrightarrow{u} 0,$$

it follows that I_n converges uniformly to $\Gamma(x)$. Therefore $\Gamma(x)$ is continuous on $[a,b]$ for any $0 < a < b < \infty$, and hence on $(0,\infty)$. By a similar procedure we can also show that its derivatives $\Gamma'(x) = \int_0^\infty e^{-t}t^{x-1}\log t dt$, $\Gamma''(x) = \int_n^\infty e^{-t}t^{x-1}(\log t)^2 dt, \cdots$ are all continuous on $(0,\infty)$.

5.3 $\Gamma\left(n+\frac{1}{2}\right) = \left(n-\frac{1}{2}\right)\cdots\left(\frac{1}{2}\right)\Gamma\left(\frac{1}{2}\right) = \dfrac{(2n)!}{n!2^{2n}}\Gamma\left(\frac{1}{2}\right)$. From Exercise 5.2 we know that $\Gamma\left(\frac{1}{2}\right) = \sqrt{\pi}$.

5.5 Use the integral definition of the gamma function to obtain

$$2^{2x-1}\Gamma(x)\Gamma\left(x+\frac{1}{2}\right) = 4\int_0^\infty\int_0^\infty e^{-(\alpha^2+\beta^2)}(2\alpha\beta)^{2x-1}(\alpha+\beta)d\alpha d\beta,$$

then change the variables of integration to $\xi = \alpha^2+\beta^2$, $\eta = 2\alpha\beta$ to arrive at the desired formula.

5.7 Apply the ratio test.

5.9 Differentiate Equation (5.12) and multiply by x.

5.11 Substitute $\nu = -1/2$ into the identity and use Exercise 5.8.

5.17 Substitute directly into Bessel's equation. Note that, whereas $J_n(x)$ is bounded at $x = 0$, $y_n(x)$ is not. Hence the two functions cannot be linearly dependent.

5.25 The definition of I_ν, as given in Exercise 5.22, extends to negative values of ν. Equation (5.18) is invariant under a change of sign of ν, hence it is satisfied by both I_ν and $I_{-\nu}$.

5.27 Follows from the bounds on the sine and cosine functions.

5.29 Applying Parseval's relation to Equations (5.22) and (5.23), we obtain

$$\int_{-\pi}^{\pi}\cos^2(x\sin\theta)d\theta = 2\pi J_0^2(x) + 4\pi\sum_{m=1}^\infty J_{2m}^2(x)$$

$$\int_{-\pi}^{\pi}\sin^2(x\sin\theta)d\theta = 4\pi\sum_{m=1}^\infty J_{2m-1}^2(x).$$

By adding these two equations we arrive at the desired identity.

5.31 Apply Lemma 3.7 to Equations (5.24) and (5.25).

5.33 (a) $\langle 1, J_0(\mu_k x)\rangle_x = \int_0^b J_0(\mu_k x)x dx = \dfrac{b}{\mu_k}J_1(\mu_k b), \|J_0(\mu_k x)\|_x^2 = \dfrac{b^2}{2}J_1^2(\mu_k b)$.

Therefore

$$1 = \frac{2}{b}\sum_{k=1}^{\infty}\frac{1}{\mu_k J_1(\mu_k b)}J_0(\mu_k x).$$

(c) $\left\langle x^2, J_0(\mu_k x)\right\rangle_x = \left(\dfrac{b^3}{\mu_k} - \dfrac{4b}{\mu_k^3}\right)J_1(\mu_k b)$. Hence

$$x^2 = \frac{2}{b}\sum_{k=1}^{\infty}\frac{\mu_k^2 b^2 - 4}{\mu_k^3 J_1(\mu_k b)}J_0(\mu_k x).$$

(e) $\langle f, J_0(\mu_k x)\rangle_x = \displaystyle\int_0^{b/2} J_0(\mu_k x)x dx = \frac{b}{2\mu_k}J_1(\mu_k b/2)$. Hence

$$f(x) = \frac{1}{b}\sum_{k=1}^{\infty}\frac{J_1(\mu_k b/2)}{\mu_k J_1^2(\mu_k b)}J_0(\mu_k x).$$

5.35 From Exercises 5.13 and 5.14(a) we have $\langle x, J_1(\mu_k x)\rangle_x = \int_0^1 J_1(\mu_k x)x^2 dx = -J_0(\mu_k)/\mu_k = J_2(\mu_k)/\mu_k$, and, from Equation (5.34), $\|J_1(\mu_k x)\|_x^2 = \frac{1}{2}J_2^2(\mu_k)$. Therefore

$$x = 2\sum_{k=1}^{\infty}\frac{1}{\mu_k J_2(\mu_k)}J_1(\mu_k x), \qquad 0 < x < 1.$$

5.37 Using the results of Exercises 5.13 and 5.14(a),

$$\begin{aligned}\langle f, J_1(\mu_k x)\rangle_x &= \int_0^1 x^2 J_1(\mu_k x)dx = \frac{1}{\mu_{k^3}}\left[2\mu_k J_1(\mu_k) - \mu_k^2 J_0(\mu_k)\right]\\ &= \frac{1}{\mu_k}J_2(\mu_k).\end{aligned}$$

Bessel's equation also implies

$$\|J_1(\mu_k x)\|_x^2 = 2[J_1'(2\mu_k)]^2 + \frac{1}{2\mu_k^2}(4\mu_k^2 - 1)J_1^2(2\mu_k) = \frac{4\mu_k^2 - 1}{2\mu_k^2}J_1^2(2\mu_k).$$

Consequently,

$$f(x) = 2\sum_{k=1}^{\infty}\frac{\mu_k J_2(\mu_k)}{(4\mu_k^2 - 1)J_1^2(2\mu_k)}J_1(\mu_k x), \qquad 0 < x < 2.$$

This representation is not pointwise. At $x = 1$, $f(1) = 1$ whereas the right-hand side is $\frac{1}{2}[f(1^+) + f(1^-)] = \frac{1}{2}$.

5.39 Assuming $u(r,t) = v(r)w(t)$ leads to

$$\frac{w'}{kw} = \frac{1}{v}\left(v'' + \frac{1}{r}v'\right) = -\mu^2.$$

Solve these two equations and apply the boundary condition to obtain the desired representation for u.

5.41 Use separation of variables to conclude that

$$u(r,t) = \sum_{k=1}^{\infty} J_0(\mu_k r)[a_k \cos \mu_k ct + b_k \sin \mu_k ct),$$

$$a_k = \frac{2}{R^2 J_1^2(\mu_k R)} \int_0^R f(r) J_0(\mu_k r) r dr,$$

$$b_k = \frac{2}{c\mu_k R^2 J_1^2(\mu_k R)} \int_0^R g(r) J_0(\mu_k r) r dr.$$

Chapter 6

6.1 (a) $\hat{f}(\xi) = \frac{2}{\xi^2}(1 - \cos \xi)$. (c) $\hat{f}(\xi) = \frac{1}{i\xi}\left(1 - e^{-i\xi}\right)$.

6.3 For any fixed point $\xi \in J$, let ξ_n be a sequence in J which converges to ξ. Because

$$|F(\xi_n) - F(\xi)| \leq \int_I |\varphi(x,\xi_n) - \varphi(x,\xi)|\, dx,$$

and $|\varphi(x,\xi_n) - \varphi(x,\xi)| \leq 2g(x) \in \mathcal{L}^1(I)$, we can apply Theorem 6.4 to the sequence of functions $\varphi_n(x) = \varphi(x,\xi_n) - \varphi(x,\xi)$ to conclude that

$$\begin{aligned}\lim_{n\to\infty} |F(\xi_n) - F(\xi)| &\leq \lim_{n\to\infty} \int_I |\varphi(x,\xi_n) - \varphi(x,\xi)|\, dx \\ &= \int_I \lim_{n\to\infty} |\varphi(x,\xi_n) - \varphi(x,\xi)|\, dx = 0.\end{aligned}$$

6.5 Suppose $\xi \in J$, and let $\xi_n \to \xi$. Define

$$\psi_n(x,\xi) = \frac{\varphi(x,\xi_n) - \varphi(x,\xi)}{\xi_n - \xi},$$

then $\psi_n(x,\xi) \to \varphi_\xi(x,\xi)$ pointwise. ψ_n is integrable on I and, by the mean value theorem, $\psi_n(x,\xi) = \varphi_n(x,\eta_n)$ for some η_n between ξ_n and ξ.

Therefore $|\psi_n(x,\xi)| \leq h(x)$ on $I \times J$. Now use the dominated convergence theorem to conclude that $\int_I \psi_n(x,\xi)dx \to \int_I \varphi_\xi(x,\xi)dx$. This proves

$$\frac{F(\xi_n) - F(\xi)}{\xi_n - \xi} \to \int_I \varphi_\xi(x,\xi)dx.$$

The continuity of F' follows from Exercise 6.3.

6.8 (a) 1, (b) 1/2, (c) 0.

6.9 Express the integral over (a,b) as a sum of integrals over the subintervals $(a,x_1),\ldots,(x_n,b)$. Because both f and g are smooth over each subinterval, the formula for integration by parts applies to each integral in the sum.

6.10 (a) f is even, hence $B(\xi) = 0$, $A(\xi) = 2\int_0^\pi \sin x \cos \xi x \; dx = 2\dfrac{1+\cos \pi\xi}{1-\xi^2}$,

and $f(x) = \dfrac{2}{\pi}\displaystyle\int_0^\infty \frac{1+\cos x\xi}{1-\xi^2} \cos x\xi \; d\xi.$

(c) $f(x) = \dfrac{2}{\pi}\displaystyle\int_0^\infty \frac{\xi - \sin \xi}{\xi^2} \sin x\xi \; d\xi.$

6.13 Define

$$f(x) = \begin{cases} e^{-x}\cos x, & x > 0 \\ -e^{x}\cos x, & x < 0. \end{cases}$$

Because f is odd its cosine transform is zero and

$$B(\xi) = 2\int_0^\infty e^{-x}\cos x \sin \xi x \; dx = \frac{2\xi^3}{\xi^4 + 4}.$$

Now $f(x)$ may be represented on $(-\infty,\infty)$ by the inversion formula (6.28),

$$e^{-x}\cos x = \frac{2}{\pi}\int_0^\infty \frac{\xi^3}{\xi^4+4} \sin x\xi \; d\xi.$$

Because f is not continuous at $x = 0$, this integral is not uniformly convergent.

6.15 Extend

$$f(x) = \begin{cases} 1, & 0 < x < \pi \\ 0, & x > \pi \end{cases}$$

as an odd function to $\mathbb{R}$ and show that its sine transform is $B(\xi) = 2(1-\cos \pi\xi)/\xi$.

6.17 Show that the cosine transform of f is

$$A(\xi) = 2\frac{1-\cos \xi}{\xi^2} = \frac{\sin^2(\xi/2)}{(\xi/2)^2}.$$

Express $f(x)$ as a cosine integral and evaluate the result at $x = 0$, which is a point of continuity of f.

6.19 Equation (6.31) implies that $\left\|\hat{f}\right\|^2 = \|A\|^2 + \|B\|^2 = 2\pi \|f\|^2$.

6.21 $\psi_n(x)$ decays exponentially as $|x| \to \infty$, so it belongs to $\mathcal{L}^1(\mathbb{R})$ and $\hat{\psi}$ therefore exists. From Example 6.17 we have $\hat{\psi}_0(\xi) = \sqrt{2\pi}\psi_0(\xi)$. Assuming $\hat{\psi}_n(\xi) = (-i)^n\sqrt{2\pi}\psi_n(\xi)$, we have

$$\begin{aligned}
\hat{\psi}_{n+1}(\xi) &= \mathcal{F}\left(e^{-x^2/2}H_{n+1}(x)\right)(\xi) \\
&= \mathcal{F}\left[e^{-x^2/2}(2xH_n(x) - H_n'(x))\right](\xi) \\
&= \mathcal{F}\left[x\psi_n(x) - \psi_n'(x)\right](\xi) \\
&= i\hat{\psi}_n'(\xi) - i\xi\hat{\psi}_n(\xi) \\
&= (-i)^{n+1}\sqrt{2\pi}[-\psi_n'(\xi) + \xi\psi_n(\xi)] \\
&= (-i)^{n+1}\sqrt{2\pi}\psi_{n+1}(x),
\end{aligned}$$

where we used the identity $H_{n+1}(x) = 2xH_n(x) - H_n'(x)$ and Theorem 6.15. Thus, by induction, $\hat{\psi}_n(\xi) = (-i)^n\sqrt{2\pi}\psi_n(\xi)$ is true for all $n \in \mathbb{N}_0$.

6.23 Define the integral $I(z) = \int_0^\infty e^{-b\xi^2}\cos z\xi\, d\xi$ and show that it satisfies the differential equation $I'(z) = -zI(z)/2b$, whose solution is $I(z) = I(0)e^{-z^2/4b}$, where $I(0) = \frac{1}{2}\sqrt{\pi/b}$.

6.25 The boundary condition at $x = 0$ implies $A(\lambda) = 0$ in the representation of $u(x,t)$ given by (6.39), so that u is now an odd function of x. By extending $f(x)$ as an odd function from $(0,\infty)$ to $(-\infty,\infty)$ we can see that $B(\lambda)$ is the sine transform of f and the same procedure followed in Example 6.18 leads to the desired result.

6.27 The transformed wave equation $\hat{u}_{tt}(\xi,t) = -c^2\xi^2\hat{u}(\xi,t)$ under the given initial conditions is solved by $\hat{u}(\xi,t) = \hat{f}(\xi)\cos c\xi t$. Taking the inverse Fourier transform yields the required representation of u.

Chapter 7

7.1 (a) $\dfrac{2a^2}{s^3} + \dfrac{2ab}{s^2} + \dfrac{b^2}{s}$.

(d) $\dfrac{1}{s^2+4}$.

(g) $\dfrac{2s}{(s^2-1)^2}$.

(i) $\sqrt{\pi/s}$.

7.2 (b) $2\cosh 3x - \dfrac{5}{3}\sinh 3x$.

(d) $\dfrac{1}{2}\left(1 - e^{-2x}\right)$.

(f) $2\sqrt{x/\pi}$.

7.5 $f(x) = x[H(x) - H(x-1)] + e^{1-x}H(x-1)$.

$\mathcal{L}(f)(\xi) = \dfrac{1}{s^2}\left(1 - e^{-s}\right) - \dfrac{1}{s}e^{-s} + \dfrac{1}{s+1}e^{-s}$.

7.6 (c) $H(x-3) + H(x-1)$.

7.7 If f has jump discontinuities at the points $x_1, \ldots, x_n$ then the sum $f(x_1^-) - f(x_1^+) + \cdots + f(x_n^-) - f(x_n^+)$ has to be added to the right-hand side of (7.6).

7.8 (e) $y(x) = H(x-1)\left[\dfrac{1}{2}e^{2(x-1)} - e^{x-1} + \dfrac{1}{2}\right] - e^x + e^{2x}$.

7.9 (c) $\dfrac{1}{x}\left(e^{-bx} - e^{-ax}\right)$.

7.11 (a) Write

$$\begin{aligned}\mathcal{L}(f)(s) &= \int_0^\infty f(x)e^{-sx}dx \\ &= \sum_{n=0}^\infty \int_{np}^{(n+1)p} f(x)e^{-sx}dx \\ &= \sum_{n=0}^\infty \int_0^p f(x+np)e^{-s(x+np)}dx,\end{aligned}$$

then use the equation $f(x+np) = f(x)$ to arrive at the answer.

(b) $\mathcal{L}(f) = \dfrac{1}{1-e^{-s}}\left[\dfrac{1}{s^2}(1-e^{-s}) - \dfrac{e^{-s}}{s}\right]$.

7.13 The left-hand side is the convolution of x^3 and $y(x)$. Applying Theorem 7.14 gives $3!Y(s)/s^4 = F(s)$, from which $Y(s) = s^4F(s)/6$. From Corollary 7.7 we conclude that

$$y(x) = \frac{1}{6}f^{(4)}(x) + \frac{1}{6}\mathcal{L}^{-1}[f(0^+)s^3 + f'(0^+)s^2 + f''(0^+)s + f'''(0^+)].$$

The integral expression for $f(x)$ implies that $f^{(n)}(0^+) = 0$ for $n = 0, 1, 2, 3$ (we also know from Exercise 7.12 that s^n cannot be the Laplace transform of a function in $\mathcal{E}$ for any $n \in \mathbb{N}_0$). Assuming that f is differentiable to fourth order (or that y is continuous), the solution is $y(x) = f^{(4)}(x)/6$.

7.15 $\mathcal{L}([x])(s) = \dfrac{e^{-s}}{s(1-e^{-s})}$.

7.17 $u(x,t) = H(t-x/c)\cos^2(t-x/c)$.

7.19 $u(x,t) = e^{-x/c}H(t-x/c)\sin(t-x/c)$.

7.21 $F(s) = e^{-a\sqrt{s}}/\sqrt{s}$ is analytic in the complex plane cut along the negative axis $(-\infty, 0]$. Using Cauchy's theorem, the integral along the vertical line $(\beta - i\infty, \beta + i\infty)$ can be reduced to two integrals, one along the bottom edge of the cut from left to right, and the other along the top edge from right to left. This yields

$$\begin{aligned}\mathcal{L}^{-1}(F)(x) &= \frac{1}{2\pi i}\int_{\beta-i\infty}^{\beta+i\infty} F(s)e^{sx}ds \\ &= \frac{1}{\pi}\int_0^\infty \frac{\cos a\sqrt{s}}{\sqrt{s}}e^{-sx}ds \\ &= \frac{2}{\pi}\int_0^\infty e^{-xt^2}\cos at\; dt.\end{aligned}$$

Noting that the last integral is the Fourier transform of e^{-xt^2}, and using the result of Example 6.17, we obtain the desired expression for $\mathcal{L}^{-1}(F)(x)$.

References

[1] Al-Gwaiz, M.A. and S.A. Elsanousi, *Elements of Real Analysis*, Chapman and Hall/CRC, Boca Raton, Florida, 2006.
[2] Birkhoff, G. and G.-C. Rota, *Ordinary Differential Equations*, 2nd edn., John Wiley, New York, 1969.
[3] Buck, R.C., *Advanced Calculus*, McGraw-Hill, 3rd edn., McGraw-Hill International, New York, 1978.
[4] Carslaw, H.S., *Introduction to the Theory of Fourier's Series and Integrals*, 3rd edn., Dover, New York, 1930.
[5] Churchill, R.V. and J.W. Brown, *Fourier Series and Boundary Value Problems*, 6th edn., McGraw-Hill International, New York, 2001.
[6] Coddington, E.A. and N. Levinson, *Theory of Ordinary Differential Equations*, McGraw-Hill, New York, 1955.
[7] Courant, R. and D. Hilbert, *Methods of Mathematical Physics*, vols I and II, Interscience Publishers, New York, 1953 and 1963.
[8] Courant, R. and F. John, *Introduction to Calculus and Analysis*, vol. II, John Wiley, New York, 1974.
[9] Folland, G.B., *Fourier Analysis and Its Applications*, Wadsworth, Belmont, California, 1992.
[10] González-Velasco, E.A., *Fourier Analysis and Boundary Value Problems*, Academic Press, San Diego, California, 1995.
[11] Halmos, P.R., *Finite-Dimensional Vector Spaces*, 2nd edn., Van Nostrand, Princeton, New Jersey, 1958.
[12] Ince, E.L., *Ordinary Differential Equations*, Dover, New York, 1956.
[13] John, F., *Partial Differential Equations*, 4th edn., Springer, New York, 1982.
[14] Rudin, W., *Principles of Mathematical Analysis*, McGraw-Hill, New York, 1964.

[15] Titchmarch, E.C., *Eigenfunction Expansions Associated with Second Order Differential Equations*, 2nd edn., Clarendon Press, Oxford, 1962.
[16] Tolstov, G.P., *Fourier Series*, Dover, New York, 1962.
[17] Watson, G.N., *A Treatise on the Theory of Bessel Functions*, 2nd edition, Cambridge University Press, Cambridge, U.K. 1944.
[18] Zettl, A., *Sturm-Liouville Theory*, vol. 121, American Mathematical Society, Providence, Rhode Island, 2005.

Notation

$\mathbb{N} = \{1, 2, 3, \ldots\}$	
$\mathbb{N}_0 = \{0, 1, 2, 3, \ldots\}$	
$\mathbb{Z} = \{\ldots, -2, -1, 0, 1, 2, \ldots\}$	
$\mathbb{Q}$	rational numbers
$\mathbb{R}$	real numbers
$\mathbb{C}$	complex numbers
$\mathbb{F}$	$\mathbb{R}$ or $\mathbb{C}$, 1
$C(I),\ C^0(I)$	set of continuous functions on the interval I, 3,5
$C^k(I)$	functions on I which have continuous derivatives up to order k, 5
$C^\infty(I)$	functions on I which have continuous derivatives of all orders, 5
$\mathcal{L}^1(I)$	Lebesgue integrable functions on the interval I, 186
$\mathcal{L}^2(I)$	Lebesgue square-integrable functions on I, 15
$\mathcal{L}^2_\rho(I)$	Lebesgue square-integrable functions on I with respect to the weight function ρ, 18
$\mathcal{E}$	locally integrable functions on $[0, \infty)$ of exponential growth at ∞, 221
$\langle \cdot, \cdot \rangle$	inner product in $\mathcal{L}^2$, 6,14
$\langle \cdot, \cdot \rangle_\rho$	inner product in $\mathcal{L}^2_\rho$, 18
$\Vert \cdot \Vert$	$\mathcal{L}^2$ norm, 14
$\Vert \cdot \Vert_\rho$	$\mathcal{L}^2_\rho$ norm, 18
$f_n \to f$	the sequence of functions f_n converges pointwise to f, 20-21
$f_n \stackrel{u}{\to} f$	f_n converges uniformly to f, 22,23
$f_n \stackrel{\mathcal{L}^2}{\to} f$	f_n converges to f in $\mathcal{L}^2$, 31-32
$\hat{f} = \mathcal{F}(f)$	Fourier transform of f, 187

$F = \mathcal{L}(f)$	Laplace transform of f, 223
$f * g$	convolution of f and g, 215
$f^{\pm}$	positive and negative parts of the function f, 111
L'	adjoint of the differential operator L, 55-56
L^*	formal adjoint of L, 57
Δ	Laplacian operator, 119
$W(f, g)$	Wronskian of f and g, 45
$G(x, \xi)$	Green's function, 68-69
$D_n(\alpha)$	Dirichlet's kernel, 106
erf	error function, 160
P_n	Legendre polynomial of order n, 133
Q_n	Legendre function of order n, 134
H_n	Hermite polynomial of order n, 141
L_n	Laguerre polynomial of order n, 146
J_ν	Bessel function of the first kind of order ν, 162
Y_ν	Bessel function of the second kind of order ν, 169
$I_\nu,\ K_\nu$	modified Bessel functions of order ν, 171
Γ	gamma function, 157
H	Heaviside function, 222

Index

Printed in the United States of America